SEMICONDUCTORS AND SEMIMETALS

VOLUME 21

Hydrogenated Amorphous Silicon

Part A

Preparation and Structure

Semiconductors and Semimetals

A Treatise

Edited by R. K. WILLARDSON

WILLARDSON CONSULTING
SPOKANE, WASHINGTON

ALBERT C. BEER

BATTELLE COLUMBUS LABORATORIES
COLUMBUS, OHIO

Willardson, Robert. K.

SEMICONDUCTORS AND SEMIMETALS

VOLUME 21
Hydrogenated Amorphous Silicon
Part A
Preparation and Structure

Volume Editor
JACQUES I. PANKOVE

RCA/DAVID SARNOFF RESEARCH CENTER
PRINCETON, NEW JERSEY

1984

ACADEMIC PRESS, INC.
(Harcourt Brace Jovanovich, Publishers)

Orlando   San Diego   San Francisco   New York   London
Toronto   Montreal   Sydney   Tokyo   São Paulo

ACADEMIC PRESS, INC.
Orlando, Florida 32887

United Kingdom Edition published by
ACADEMIC PRESS, INC. (LONDON) LTD.
24/28 Oval Road, London NW1 7DX

Library of Congress Cataloging in Publication Data
Main entry under title:

Semiconductors and semimetals.

    Includes bibliographical references and indexes.
    Contents: v. 1-2. Physics of III-V compounds -- v. 3.
Optical properties of III-V compounds -- [etc.] -- v. 21.
Hydrogenated amorphous silicon. pt. A. Preparation and
structure / edited by Jacques I. Pankove.
    1. Semiconductors--Collected works. 2. Semimetals--
Collected works. I. Willardson, Robert K. II. Beer,
Albert C.
QC610.9.S47   537.6'22   65-26048
ISBN 0-12-752121-6 (v.21A)

Contents

Chapter 1 Introduction 1

Jacques I. Pankove

PREPARATION

Chapter 2 Glow Discharge

Masataka Hirose

Chapter 3 dc Glow Discharge

Yoshiyuki Uchida

Chapter 4   Sputtering

T. D. Moustakas

Chapter 5   Ionized-Cluster Beam Deposition

Isao Yamada

Chapter 6   Chemical Vapor Deposition

Masataka Hirose

Chapter 7   Homogeneous Chemical Vapor Deposition

Bruce A. Scott

CHARACTERISTICS OF SILANE PLASMA

Chapter 8   Chemical Reactions in Plasma Deposition

Frank J. Kampas

Chapter 9   Plasma Kinetics

Paul A. Longeway

Chapter 10   Diagnostics of Silane Glow Discharges
Using Probes and Mass Spectroscopy

Herbert A. Weakliem

STRUCTURE

List of Contributors

Numbers in parentheses indicate the pages on which the authors' contributions begin.

DAVID ADLER, *Department of Electrical Engineering and Computer Science and Center for Materials Science and Engineering, Massachusetts Institute of Technology, Cambridge, Massachusetts 02139* (291)

A. CHENEVAS-PAULE, *Centre d'Etudes Nucléaires de Grenoble, Laboratoire d'Electronique et de Technologie de l'Informatique, 38041 Grenoble, France* (247)

LESTER GUTTMAN, *Materials Science and Technology Division, Argonne National Laboratory, Argonne, Illinois 60439* (225)

MASATAKA HIROSE, *Department of Electrical Engineering, Hiroshima University, Higashihiroshima 724, Japan* (9, 109)

FRANK J. KAMPAS, *Division of Metallurgy and Materials Science, Brookhaven National Laboratory, Upton, New York 11973* (153)

PAUL A. LONGEWAY, *RCA Laboratories, Princeton, New Jersey 08540* (179)

S. MINOMURA, *Institute for Solid State Physics, University of Tokyo-Roppongi, Tokyo 106, Japan* (273)

T. D. MOUSTAKAS, *Corporate Research Laboratories, Exxon Research and Engineering Company, Annandale, New Jersey 07036* (55)

JACQUES I. PANKOVE, *RCA/David Sarnoff Research Center, Princeton, New Jersey 08540* (1)

BRUCE A. SCOTT, *IBM Thomas J. Watson Research Center, Yorktown Heights, New York 10598* (123)

YOSHIYUKI UCHIDA, *Semiconductor Laboratories, Fuji Electric, Corporate Research and Development, Ltd., Nagasaka, Yokosuka City, Japan* (41)

HERBERT A. WEAKLIEM, *RCA Laboratories, Princeton, New Jersey 08540* (195)

ISAO YAMADA, *Ion Beam Engineering Experimental Laboratory, Kyoto University, Sakyo, Kyoto 606, Japan* (83)

xi

Foreword

This book represents a departure from the usual format of "Semiconductors and Semimetals" because it is a part of a four-volume miniseries devoted entirely to hydrogenated amorphous silicon (a-Si:H). In addition, this group of books—Parts A–D of Volume 21—has been organized by a guest editor, Dr. J. I. Pankove, an internationally recognized authority on this subject. He has assembled most of the who's who in this field as authors of the many chapters. It is especially fortunate that Dr. Pankove, who has made important original contributions to our understanding of a-Si:H, has been able to devote the time and effort necessary to produce this valuable addition to our series. In the past decade, a-Si:H has developed into an important family of semiconductors. In hydrogenated amorphous silicon alloys with germanium, the energy gap decreases with increasing germanium content, while in alloys with increasing carbon content the energy gap increases. Although many applications are still under development, efficient solar cells for calculators have been commercial for some time.

In Volume 21, Part A, the preparation of a-Si:H by rf and dc glow discharges, sputtering, ion-cluster beam, CVD, and homo-CVD techniques is discussed along with the characteristics of the silane plasma and the resultant atomic and electronic structure and characteristics.

The optical properties of this new family of semiconductors are the subject of Volume 21, Part B. Phenomena discussed include the absorption edge, defect states, vibrational spectra, electroreflectance and electroabsorption, Raman scattering, luminescence, photoconductivity, photoemission, relaxation processes, and metastable effects.

Volume 21, Part C is concerned with electronic and transport properties, including investigative techniques employing field effect, capacitance and deep level transient spectroscopy, nuclear and optically detected magnetic resonance, and electron spin resonance. Parameters and phenomena considered include electron densities, carrier mobilities and diffusion lengths, densities of states, surface effects, and the Staebler–Wronski effect.

The last volume of this miniseries, 21, Part D, covers device applications, including solar cells, electrophotography, image pickup tubes, field effect transistors (FETs) and FET-addressed liquid crystal display panels, solid state image sensors, charge-coupled devices, optical recording, visible light emitting diodes, fast modulators and detectors, hybrid structures, and memory switching.

R. K. WILLARDSON
ALBERT C. BEER

Preface

Hydrogenated amorphous silicon, a new form of a common element, is a semiconductor that has come of age. Its scientific attractions include a continuously adjustable band gap, a usable carrier lifetime and diffusion length, efficient optical transitions, and the capability of employing either n- or p-type dopants.

Furthermore, it can be fabricated very easily as a thin film by a technology that not only inherently escapes the expense of crystal perfection but also requires significantly smaller amounts of raw materials.

The discovery of a new material endowed with wondrous possibilities for very economical practical applications naturally attracts many researchers who invariably provide new insights and further vision. Their meditation and experimentation build up rapidly and lead to a prolific information flow in journals and conference proceedings.

The initial cross-fertilization generates an overload of data; books are written that attempt to digest specialized aspects of the field with state-of-the-art knowledge that often becomes obsolete by the time the books are published a year or two later.

We have attempted to provide this book with a lasting quality by emphasizing tutorial aspects. The newcomer to this field will not only learn about the properties of hydrogenated amorphous silicon but also how and why they are measured.

In most chapters, a brief historical review depicts the evolution of relevant concepts. The state of the art emerges, and a bridge to future developments guides the reader toward what still needs to be done. The abundant references should be a valuable resource for the future specialist.

We hope that this tutorial approach by seasoned experts satisfies the needs of at least one generation of new researchers.

JACQUES I. PANKOVE

CHAPTER 1

Introduction

Jacques I. Pankove

RCA/DAVID SARNOFF RESEARCH CENTER
PRINCETON, NEW JERSEY

The purpose of this introductory chapter is to outline in simple terms the contents of this volume. Occasionally, reference will be made to concepts that are relevant to the topic but outside the main scope of the authors.

There are many ways to prepare hydrogenated amorphous silicon (a-Si:H). The most common method is the glow-discharge decomposition of silane, wherein an electric field (dc, ac, rf) is used to produce a plasma containing ions and other reactive species, which condense on a heated substrate (200–400°C, typically) to form an amorphous solid still rich in hydrogen. Many reactions can occur at the substrate and, as will be shown by P. A. Longeway (Chapter 9), for dc plasma, some of the reactions may involve the insertion of an SiH_3 molecule that can subsequently lose all or some of its hydrogen.

There are many variants of the glow-discharge method. In the rf method discussed by M. Hirose (Chapter 2), the power is coupled inductively or capacitively using external electrodes, or capacitively with internal electrodes; an additional dc transverse electric field can be used to stabilize the discharge or a dc magnetic field can be used to confine the plasma, increasing its density and minimizing contamination due to bombardment of the walls. A dc discharge can be used, as discussed by Y. Uchida (Chapter 3), whereby different deposition processes occur on the anode and the cathode. Ion bombardment of the film is reduced by a "proximity electrode" that sets an equipotential plane at the same potential as the substrate but away from the surface of the film. The glow discharge also may be fed at the power-line frequency, in which case it is called an ac glow discharge.

Sputtering (or more accurately, "reactive sputtering") is described by T. D. Moustakas (Chapter 4). Sputtering is a well-developed industrial process capable of fast deposition. Sputtering also is a glow-discharge technique, but in this case, the source of silicon is a solid silicon target that is bombarded by ~ 1-keV ions (usually argon). Si atoms are thus sputtered and transported through the plasma to a heated substrate. A controlled amount of hydrogen added to the sputtering gas is atomized and reacts with the

1

sputtered Si. In the case of sputtering and glow-discharge depositions, a two-dimensional array of parameters can be varied simultaneously during one deposition (e.g., doping and substrate temperature or target-to-substrate spacing) (Hanak *et al.*, 1979; Carlson, 1980).

The evaporation of silicon or sputtering without hydrogen produces an amorphous film that can be subsequently hydrogenated by exposure to atomic hydrogen (Sol *et al.*, 1980). Molecular beam techniques using effusion cells can be used to deposit Si and H simultaneously. When Si is vaporized by a pulsed Nd:YAG laser in a hydrogen ambient, a-Si:H deposits at an extraordinarily high rate of 10^6 Å sec^{-1} (Hanabusa and Suzuki, 1981).

The ion-cluster beam deposition method, described by I. Yamada (Chapter 5), is designed to condense the Si vapor into aggregates whose size is on the order of 100–1000 atoms; these clusters are ionized and accelerated toward the substrate, where their kinetic energy is controlled to be in the range of 0.1–1 eV per atom. This is much lower than the energy per atom produced in either the glow-discharge or sputtering processes. A partial pressure of hydrogen allows H-incorporation during the deposition.

The pyrolytic decomposition of silane at 450°C produces (at a slow rate) an amorphous film that contains a small concentration (~ 2 at. %) of hydrogen. This technique is known as chemical vapor deposition (CVD) and is described by M. Hirose (Chapter 6). Above ~ 500°C, hydrogen evolves rapidly from the film. At ~ 600°C a microcrystalline layer is obtained, and above ~ 700°C the material becomes polycrystalline. When disilane is used as the source of Si, the CVD deposition rate is much faster and can proceed at much lower temperatures (e.g., 300°C), thereby reducing the amount of hydrogen that is lost.

If a cooled substrate is inserted into a CVD reactor with the gas heated to ~ 700°C, the material condensing on the substrate is a-Si:H having a high concentration of hydrogen (up to 40%), a low concentration of dangling bonds, and apparently few weakly bonded Si atoms. This technique is called "HomoCVD," because the substrate samples a homogeneous chemical reaction. It is described by its inventor, B. A. Scott (Chapter 7).

Silane may be directly decomposed by photons having sufficient energy; this is ultraviolet photolysis (Perkins *et al.*, 1979). Photolysis of silane also can be achieved in the infrared by resonant coupling to the vibrational mode of a Si–H bond using the 10.59-μm emission line from a Co_2 laser (Hanabusa *et al.*, 1979). Although photolysis should minimize contamination of the a-Si:H obtained, there has been no extensive study of films made by this technique.

Doping in most of the preceding techniques is obtained by adding either phosphine or diborane to silane to produce *n*- or *p*-type a-Si:H, respectively.

In the case of sputtering, the impurity can be incorporated into the solid target or supplied from the gas phase. Sometimes fluorine is used instead of, or in addition to, hydrogen to passivate the dangling bonds.

In hydrogenated amorphous alloys of germanium and silicon the energy gap decreases with increasing germanium content (Hauschildt *et al.*, 1980). Conversely, hydrogenated amorphous alloys of silicon and carbon result in a larger bandgap as the carbon concentration is increased (Sussman and Ogden, 1981). a-$B_{1-x}H_x$ (Pankove and Hough, 1979) and its alloy with a-Si:H (Tsai, 1979) also have been explored.

The plasma is comprised of numerous ions, radicals, and neutral species as well as electrons. Their interactions are reviewed by F. A. Kampas (Chapter 8). Inside a plasma, a rich set of chemical reactions takes place as molecules are broken and the pieces reassembled into various structures, many of which are short-lived. SiH_4 is broken into SiH_3, SiH_2, SiH, and Si, but these fragments can combine to form Si_2H_6, Si_3H_8, $(SiH_2)_n$, and so on. It was long thought that SiH_2 may be a precursor to the formation of a-Si:H films by its attachment to dangling bonds or by insertion reactions at the surface of the film with a subsequent loss of hydrogen. However, recent experiments by Longeway (Chapter 9) show that the addition of NO (an SiH_3 scavenger) stops film growth, suggesting that SiH_3 may be an important intermediate that binds to dangling bonds at the surface of the film before losing its hydrogen.

Powerful tools are available to characterize a plasma. They are (1) emission spectroscopy, which utilizes the various components in the luminescent glow that have been identified (see Chapters 2 and 8); (2) mass spectroscopy, which can identify most of the masses (both ions and neutrals) present in the plasma; and (3) Langmuir probe measurements, from which plasma density, plasma potential, and electron energy distribution can be calculated. The last two methods are discussed by H. A. Weakliem (Chapter 10).

A knowledge of the structure of a-Si:H is important before one can proceed with an intelligent interpretation of optical and electrical properties of the material and before one can design viable devices. L. Guttman (Chapter 11) presents the theorist's perception of how various reasonable atomic arrangements can be tested, how the models lead to a description of the electronic structure in amorphous silicon, and how the addition of hydrogen affects this description. The type of clustering considered and the assumption of voids in the material have important consequences on the properties of the film.

The structure of a-Si:H can be probed experimentally [described by A. Chenevas-Paule (Chapter 12)] using high-resolution transmission electron microscopy to image two-dimensional features on the order of a few angstroms or scanning electron microscopy, which gives a three-dimensional view on a coarser scale. Neutron scattering can probe the homogeneity of a material down to a fraction of an angstrom sensitivity and the scattering can be hydrogen-specific. Using a combination of these techniques, a new tentative picture emerges for sputtered posthydrogenated a-Si:H. According to this new study (limited to one type of sample), hydrogen forms a lining that "decorates" microtubules about 120 Å long and about 30 Å apart. Earlier structural studies (Knights, 1980) have shown that it is possible to obtain a columnar structure that sometimes contains large nodules. More discussion of structure in sputtered films will be found in Chapter 4. It is evident that much more work is needed to correlate the details of the film structure with those deposition parameters that can be readily controlled.

Other techniques for studying the structure are (1) x-ray diffraction, which can yield (in thick films) a radial distribution function of Si–Si bond lengths and qualitative information about bond angles; and (2) Raman scattering, which can probe the presence of Si–H bonds and Si–lattice vibrational modes. S. Minomura (Chapter 13) has examined a variety of samples of a-Si:H using these two techniques while changing the hydrostatic pressure up to 180 kbar. The changes are interpreted in terms of modified bond lengths and bond angles. Electrical measurements reveal that high pressure can induce a phase transition to the metallic state, and even superconductivity can be obtained in a-Si:H.

Having considered the details of the many possible bonding configurations in a strained amorphous structure, D. Adler (Chapter 14) explores the resulting distribution of energy levels. He shows how hydrogen, with a binding energy to Si stronger than that of silicon to silicon, causes the gap energy to increase and how disorder can generate a tailing of the states into the energy gap. An amazingly large variety of defects, mostly strain related, can be frozen into the structure during deposition, forming localized states that lie deep inside the energy gap. Among these deep states the most important are the dangling bonds that act as efficient recombination centers and therefore as lifetime killers. Adding impurities, such as phosphorus, boron, or oxygen, further complicates the variety of possible defects.

Recently, Abeles and Tiedje (1983) made "superlattices" consisting of 10–1000-Å-thick layers of a-Si:H alternating with layers of a-Ge:H; a-Si$_{1-x}$C$_x$; or a-Si:N$_x$. The films were deposited by a glow discharge technique in which the composition of the reactive gases was changed periodically. Atomically sharp interfaces were obtained by exchanging the gases in the plasma reactor in a time that was short compared to the time it takes to grow a monolayer. The sharpness of the interfaces and the smoothness and uniformity of the layers was demonstrated by x-ray diffraction measurements of the superlattice. The films exhibit a low density of gap states, presumably because of the ability of hydrogen to passivate interfacial as well as bulk defects. The ability to make low defect density superlattices with atomically sharp interfaces opens up the possibility of synthesizing a new class of amorphous semiconductors exhibiting a diverse range of electronic behavior and potential device applications.

REFERENCES

Abeles, B., and Tiedje, T. (1983). *Phys. Rev. Lett.* **51**, 2003.
Carlson, D. E. (1980). *In* "Polycrystalline and Amorphous Thin Films and Devices" (L. Kazmerski, ed.), p. 175. Academic Press, New York.
Hanabusa, M., and Suzuki, M. (1981). *Appl. Phys. Lett.* **39**, 431.
Hanabusa, M., Namiki, A., and Yoshihara, K. (1979). *Appl. Phys. Lett.* **35**, 626.
Hanak, J. J., Korsun, V., and Pellicane, J. P. (1979). *In* "Proceedings of the Second European Communities Photovoltaic Solar Energy Conference, " p. 270. Reidel Publ., Dordrecht, Netherlands.
Hauschildt, D., Fischer, R., and Fuhs, W. (1980). *Phys. Status Solidi B* **102**, 563.
Knights, J. C. (1980). *Sol. Cells* **2**, 409.
Pankove, J. I., and Hough, W. V. (1979). *J. Appl. Phys.* **50**, 6018.
Perkins, G. G. A., Austin, E. R., and Lampe, W. F. (1979). *J. Am. Chem. Soc.* **101**, 1109.
Sol, N., Kaplan, D., Dieumegard, D., and Dubrevil, D. (1980). *J. Non-Cryst. Solids* **35**, 291.
Sussman, R. S., and Ogden, R. (1981). *Philos. Mag. [Part] B* **44**, 137.
Tsai, C. C. (1979). *Phys. Rev. B* **19**, 2041.

Preparation

CHAPTER 2

Glow Discharge

Masataka Hirose

DEPARTMENT OF ELECTRICAL ENGINEERING
HIROSHIMA UNIVERSITY
HIGASHIHIROSHIMA, JAPAN

I. Introduction

1. GENERAL FEATURES OF GLOW DISCHARGE

Plasma deposition and etching are potentially important techniques for fabricating novel devices and synthesizing a new class of electronic materials as well. The growing interest in the plasma processing motivates recent extensive studies on plasma chemistry, and numerous types of plasma deposition systems and etching reactors are now being used in the electronic industries to produce integrated circuits. Recent success in the device applications of amorphous hydrogenated silicon (a-Si:H) is also based on the plasma deposition technique. Device quality a-Si:H films have been prepared so far by empirical control of the chemical reactions that proceed in a silane plasma and on the substrate surface. Therefore it will be useful to describe the fundamental aspects of the glow discharge phenomena that occur in a typical reactor.

a. Plasma Potential and Self-Bias

A typical plasma reactor for plasma etching and deposition is a flat-bed system, in which the discharge is generated between two parallel disk electrodes (see the inset of Fig. 1). The plasma potential V_p is the potential of the glowing part of the discharge and is most positive anywhere in the reactor. The potential V_p is at least as large as the first ionization potential of the gas, with respect to the grounded electrode in contact with the glow discharge, and it is regarded as the reference potential of the system. It should be noted that the ground potential in a flat-bed reactor is always negative with respect to the plasma, and hence the substrate surface on the grounded electrode should more or less suffer the positive ion bombardment. The potential on the powered electrode is also negative with respect to the plasma potential (Vossen, 1979). This is explained as follows: The large difference in mobility between electrons and ions in a glow discharge results in the static I–V characteristic of the plasma being similar to a leaky diode, as shown in Fig. 1. An rf voltage applied to this type of load induces a large electron current toward the powered electrode during one-half of the cycle and a small ion current on the other half of the cycle. As a result, the capacitor connected to the rf power source is negatively charged to develop the average self-bias V_{dc}, and finally the net electron current induced by the positive voltage exceeding V_p on one-half of the cycle becomes equal to the net ion current on the other half of the voltage cycle, during which the electrode potential is lower than V_p. The average potential distribution in a reactor is shown in Fig. 2. The sheath potential V_s, being equal to $V_{dc} + V_p$

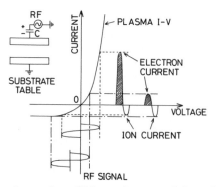

FIG. 1. Development of a negative self-bias on the powered electrode in contact with an rf glow discharge. (From Vossen, 1979.)

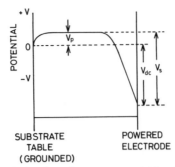

FIG. 2. Spatial distribution of average potential in a flat-bed reactor.

on the powered electrode, accelerates ions that enter the sheath, and the electrode surface is bombarded. In a standard flat-bed reactor powered with an rf source, the substrate is placed on the grounded electrode to reduce the influence of the ion bombardment. However, the plasma potential is still important to determine the energy of ions impinging onto the substrate surface. The plasma potential could, in principle, be measured by an electric probe (e.g., Chen, 1965).

b. Electron Temperature

The ordinary glow discharge employed in the plasma deposition is weakly ionized plasma, because the degree of ionization, which refers to the ratio of electron density to gas molecule density, is at most $\sim 10^{-4}$. Therefore, the electron and ion densities in the plasma at a pressure of 0.1 Torr (13 Pa) might be less than $\sim 10^{11}$ cm^{-3}. The electron temperature T_e can be uniquely defined, if the electron energy distribution function is assumed to be Maxwellian and is in the range of $\sim 10^{4}$°K, while the ion and gas temperatures (about several hundred degrees kelvin) are far below T_e. Such a nonequilibrium plasma contains an appreciable amount of high-energy electrons, which encounters gas molecules to produce ions and radicals. Here the term "radicals" involves the whole chemically active neutral species.

The ionization and radical formation through the electron impact cause a significant deviation of the electron energy distribution function from the Maxwellian. A possible way to estimate the electron distribution function $f(\varepsilon)$ is to employ the electric probe technique, because a second differentiation of the current–voltage curve for the probe immersed in a plasma yields $f(\varepsilon)$ (Chen, 1965). This technique has been applied to the dc glow discharge of silane, and the result showed that the measured distribution function is significantly different from both Maxwell and Druyvestein distributions, as shown in Fig. 3 (Kocian, 1980). Most of electron-impact dissociation

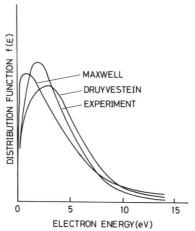

FIG. 3. An electron energy distribution function for the silane plasma obtained by the electric probe technique. (From Kocian, 1980.)

reactions in the silane plasma will occur in the electron energy range 2 – 7 eV and dissociative ionizations might take place at energies above 11 eV (Turban *et al.*, 1982). The electrons in the silane plasma can therefore give rise to the efficient formation of neutral radicals, as expected from the energy distribution function of Fig. 3.

c. Excitation Frequency

Wide varieties of excitation frequencies from dc to microwave are used for generating glow discharges. High- and low-frequency regions will be distinguished as follows: At high frequencies only electrons can follow the applied ac electric field, whereas at low frequencies ions collide with the surface of the powered electrode during one-half of the cycle and electrons on the other one-half of the cycle. The critical frequency f_c, above which ions can no longer respond to the alternating electric field, is given by

$$f_c = \mu_i E_0 / \pi l. \tag{1}$$

Here l is the interelectrode spacing, μ_i is the ion mobility, and E_0 is the amplitude of ac electric field. Since the ion mobility for various gases is in the range $\sim 10^3 - 10^4$ cm^2 V^{-1} sec^{-1}, the critical excitation frequency f_c is estimated to be $\sim 6 - 60$ kHz for typical values of $l = 5$ cm and $E = 100$ V cm^{-1}. Ions become immobile in a glow discharge generated at frequencies above ~ 1 MHz. Therefore, primary chemical reactions in an rf plasma may be initiated by electrons accelerated by the electric field.

II. Process Monitoring

Physical properties of plasma-deposited a-Si : H are sensitively influenced by discharge conditions such as pressure, gas flow rate, and power. Since numerous variants of plasma reactors, such as flat-bed systems with internal electrodes and tube reactors with external electrodes, are widely used, the deposition conditions must be optimized for the respective system. At present state, the detailed mechanism of a-Si : H growth is little understood because of the great complexity of chemical processes that proceed in the silane plasma (Kampas and Griffith, 1981), and hence the material properties are empirically controlled by finding a good combination of the deposition parameters. In situ observations of some of chemical species created in

TABLE I

OPTICAL WAVELENGTHS FOR REACTIVE SPECIES IN THE
SILANE PLASMA

Species	Emission wavelength, λ (nm)	Transition	Energy of emitting state above ground state (eV)
Si	198–199	UV 7	6.3
Si	206	UV 52	6.8
Si	212	UV 48	6.6
Si	221–222	UV 3	5.6
Si	229–230	UV 46	6.2
Si	244	UV 45	5.9
Si	251–253	UV 1	4.9
Si	263	UV 83	6.6
Si	288	UV 43	5.1
Si	299	1	5.0
Si	391	3	5.1
SiH	386–388	$A\,^2\Delta - X\,^2\Pi$	3.0
SiH	394–396	$A\,^2\Delta - X\,^2\Pi$	3.0
SiH	413–428	$A\,^2\Delta - X\,^2\Pi$	3.0
SiH$^+$	399	$A\,^1\Pi - X\,^1\Sigma$	3.1
H	434	H_γ	13.0
H	486	H_β	12.7
H	656	H_α	12.0
H$_2$	Continuum 160–500	$2s\,^3\Sigma - 2p\,^3\Sigma$	11.9
H$_2$	Many lines 368–835	$3d\,^1\Sigma - 2p\,^1\Sigma$	4.0
		$3d\,^1\Pi - 2p\,^1\Sigma$	4.0
		$3p\,^3\Pi - 2s\,^3\Sigma$ etc.	

the glow discharge will be necessary for obtaining satisfactory reproducibility of the film properties and for more general understanding of the plasma chemistry. This is the reason that in-process monitoring techniques have to be developed. Optical emission spectroscopy, mass spectrometry, and infrared absorption spectroscopy are three diagnostic techniques that have been applied to the study of the plasma reactions.

2. OPTICAL EMISSION SPECTROSCOPY (OES)

The light-emitting species in the glow discharge of silane are easily detected by a conventional spectroscopic technique, and the emission wavelength of the respective species is summarized in Table I (Kampas and Griffith, 1980). The optical emission spectra have been measured by the use of the systems, as illustrated in Fig. 4a and b (Kampas and Griffith, 1980; Hamasaki et al., 1980). In a flat-bed reactor of Fig. 4a, a quartz window is used to probe the emitted light at the end of a 3-in. port, which contains a metal screen to prevent film deposition on the window. Emitted light is incident to a monochromator with a cooled photomultiplier (R136 or R955, Hamamatsu TV Co., Ltd.). The monochromator with a smaller focal length is useful for routine monitoring of deposition conditions. In a tube reactor of Fig. 4b, rf power was supplied through two external ring electrodes 5 cm apart. A magnetic field was applied perpendicularly to the substrate surface with field lines to confine the plasma. Optical emission is collected by a mirror installed in the reactor and focused by a quartz lens either onto the edge of an optical fiber or on the entrance slit of a 25-cm monochromator with an R453 Hamamatsu photomultiplier. Phase-sensitive detection of photomultiplier output provides the low-noise spectra. Light signal processing can be done also by employing an optical multichannel analyzer (Matsuda et al., 1980). A typical emission spectrum obtained by the system in Fig. 4b is shown in Fig. 5. The emission intensities of Si, SiH, H, and H_2 are sensitively changed by varying rf power, gas flow rate, pressure, and magnetic field, because each of these parameters affects the electron density and/or the electron temperature. The emission intensity I_{ji} is related to the concentration of the emitting state radical N_j by the following equation:

$$I_{ji} = N_j P_{ji} h\nu_{ji}, \tag{2}$$

where P_{ji} is the optical transition probability from the excited state j to the ground state i and $h\nu_{ji}$ is the corresponding photon energy emitted. The concentrations thus obtained of the emissive Si (288 nm), SiH (414 nm), and H (656 nm) are represented in Fig. 6, using the values of $P_{ij} = 1.75 \times 10^8$, 1.43×10^6, and 6.41×10^7 sec^{-1}, respectively (Hirose et al., 1981).

The reactive species identified by the emission spectrum is rather limited and there is no emission from SiH_2 and SiH_3, which are generally regarded

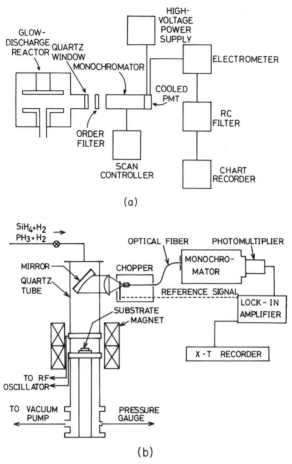

FIG. 4. (a) Block diagram of the apparatus for OES. (From Kampas and Griffith, 1980.) (b) Experimental setup for depositing a-Si:H and for optical emission spectroscopy. (From Hamasaki *et al.*, 1980.)

as the potentially important species in silicon deposition. However, it is worthwhile to note the fact that the quantity of SiH bonds incorporated in the a-Si:H matrix is proportional to the optical emission intensity of SiH (Fig. 7) and that the deposition rate measured as a function of total gas flow rate exhibits a good correlation with the product of the intensities of emissions by SiH and H radicals (Hirose, 1982a,b). This suggests the presence of some causal relationships between the optical emission spectra and the deposition kinetics, and therefore the optical emission spectroscopy of the silane plasma could be one of monitoring techniques to look at the

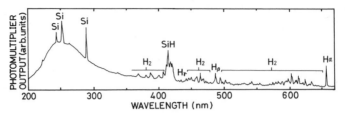

FIG. 5. Optical emission spectrum from silane glow discharge. $SiH_4/H_2 = 11\%$. (From Hamasaki *et al.*, 1980.)

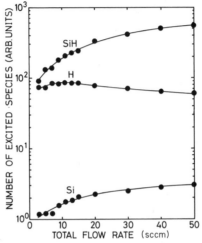

FIG. 6. Total flow-rate dependence of the radical concentrations in the silane plasma. Silane concentration is 11% in H_2. $P = 0.64$ Torr. (From Hirose, 1982a.)

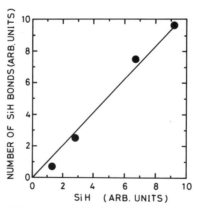

FIG. 7. Number of Si–H bonds in a-Si : H versus the emitting SiH concentration. $P = 0.64$ Torr. rf power = 20 W. $SiH_4/H_2 = 11\%$ (From Hirose, 1982b.)

state of the glow discharge. The advantage of the optical monitoring is that the plasma is not perturbed and the measurement system is simple. It is also easy to feed back the signal to the control units of a mass flowmeter, throttle valve, and matching circuits in order to keep a plasma stable.

3. MASS SPECTROMETRY

The ionic and neutral species present in the glow discharge of silane can be studied by means of mass spectrometry. A schematic of the experimental arrangement for taking mass spectra is shown in Fig. 8 (Turban *et al.*, 1980). The glow discharge is generated between two parallel internal electrodes of length 15 cm. The sampling orifice with a hole of 200 μm in diameter is perforated in a ceramic cylinder, which is chosen in order to avoid any spurious discharges that sometimes occur on a metallic sampling orifice. The resulting beam passes through the first differentially pumped stage and reaches the ionization chamber, which is evacuated by the second-stage pump. In the case of ionic measurement, the ionization chamber of the mass filter is turned off and the positive ions are directly extracted from the plasma. Any special focusing is not employed to avoid further reactions in the sheath of the orifice. When neutrals are measured, the ions in the extracted beam are magnetically deflected out of the intermediate chamber

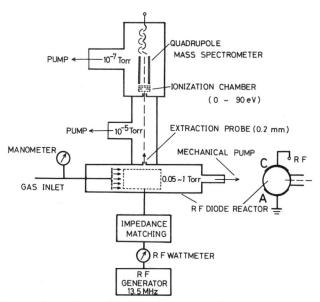

FIG. 8. The experimental setup for the mass spectrometric study. The rf power, pressure, and flow rate of 5% SiH$_4$ with H$_2$ were, respectively, 2.5–50 W, 0.05–5 Torr, and 4–42 sccm. (From Turban *et al.*, 1980.)

TABLE II

AP FOR SiH_n^+ IONS AND IP FOR SiH_n
RADICALS CREATED FROM SiH_4

Ion	Mass	AP (eV)	IP (eV)
SiH_3^+	31	12.3	8.1
SiH_2^+	30	11.9	9.5
SiH^+	29	15.3	7.4
Si^+	28	13.6	8.2

and the ionization chamber is turned on. If the electron energy for impact ionization is chosen slightly below the appearance potential (AP) of the SiH_n^+ ($n = 0, 1, 2, 3$) ions but higher than the ionization potential (IP) of the SiH_n radicals (see Table II and Fig. 9), quantitative determination of radical concentrations will, in principle, be possible. However, the sensitivity of the mass spectrometer is not sufficient for such low-energy electrons. The intense resonance lines (10.03 and 10.64 eV) from Kr produced by microwave discharge may be available as an ionization source for the SiH_n radicals, because it recently has been demonstrated that the Ar resonance lines at 11.6 and 11.8 eV can be used to ionize the radicals extracted from the CF_4 plasma (Hayashi et al., 1982). An attempt to estimate the SiH_n radical concentration has been made using the standard ionization chamber operating at 90 eV (Bourquard et al., 1981), but further details are not described here. It should be emphasized that the high reactivity of free

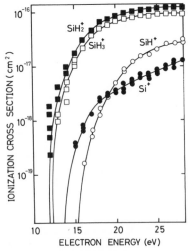

FIG. 9. Partial ionization cross sections of silane. (From Turban et al., 1982.)

radicals makes them very sensitive to collisions with the wall of the sampling orifice.

4. INFRARED ABSORPTION AND EMISSION

The optical emission spectroscopy in the visible–UV range provides no information on SiH_2, SiH_3, and SiH_4 molecules in the plasma, and the vibration–rotation transitions of SiH_m ($m = \sim 2-4$) species have either been observed or are predicted in the infrared regime. The ν_3 band of SiH_4 centered at 2189 cm^{-1} has been extensively studied so far (Dang-Nue *et al.*, 1974), and strong vibrational absorptions have been observed for the radicals SiH, SiH_2, and SiH_3 in the electronic ground states created by means of photolysis of silane in inert gases (Milligan and Jacox, 1970). They observed the characteristic absorption bands appearing at 1950–2050 cm^{-1} for SiH, 2020 cm^{-1} for SiH_2, 926 cm^{-1} for SiH_3, and 2189 cm^{-1} for SiH_4. More recently, high-resolution absorption and emission spectroscopy of the rf glow discharge of silane has been attempted in the 1800–2300 cm^{-1} range by the use of a modified Connes-type interferometer with $\sim$ 1-m-long cell (Knights *et al.*, 1982). The result showed that emission from the ground electronic state of the free radical SiH is observed and the relative ground-state SiH concentration with respect to SiH_4 is estimated to be $\sim 5 \times 10^{-3}$. There is no observed emission from SiH_2 or SiH_3, because of the lack of

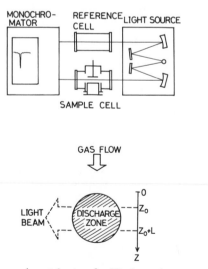

FIG. 10. Schematic of experimental setup for IR absorption spectroscopy and discharge zone in the sample cell. The flow rate of 11% SiH_4 in H_2 at a pressure of 10 Torr was varied from 30 to 50 sccm. (From Hamasaki *et al.*, 1982.)

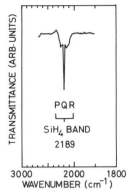

FIG. 11. Infrared absorption spectrum of SiH$_4$. (From Hamasaki *et al.*, 1982.)

symmetry for both species, which leads to the splitting of each emission line into $(2J + 1)$ components.

A simple way to measure the infrared absorption of the silane plasma is to employ a conventional dual-beam infrared spectrometer. The discharge and reference cells with the same geometry were placed in the dual beam as shown in Fig. 10 (top). A zone of cylindrical plasma (radius = 3 cm and length = 0.8 cm) was sustained by the dc glow discharge between the parallel electrodes. In Fig. 10 (bottom), the discharge zone through which the infrared light beam is transmitted is sketched, where the direction of gas flow is a z direction. In this system the intensity of the v_3 spectrum of silane, which consists of the central line of Q branch at 2189 cm^{-1} and the broad satellites of P and R branches (see Fig. 11), is observable and strongly dependent on the discharge power. The intensity ratio of the v_3 spectrum for the discharge cell to the reference cell can yield the dissociation efficiency of silane R_D. The dc power dependence of measured R_D (Fig. 12) is empirically fitted to the following equation:

$$R_D = 100 \times [1 - \exp(-W/W_0)] \quad \text{percent.} \tag{3}$$

Here W is the dc power and W_0 is the constant depending on the gas flow rate (Hamasaki *et al.*, 1982). The power dependence of R_D has been calculated by the continuity equation for the silane concentration $N(z)$ in the discharge zone along the gas flow. The continuity equation is given by

$$u\frac{\partial N(z)}{\partial z} = -k_1 N(z), \tag{4}$$

where u is the mean linear flow velocity and k_1 is the decomposition rate of silane per unit length. The diffusion effect of silane gas is estimated to be

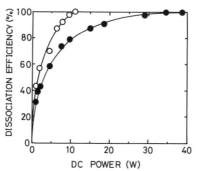

FIG. 12. Dissociation efficiency of SiH_4 gas as a function of dc power at different flow rates. $P = 10$ Torr. Flow rate (sccm): O, 30; ●, 50. (From Hamasaki *et al.*, 1982.)

negligible. Solving Eq. (4), the silane dissociation efficiency R_D in percent is obtained as

$$R_D = 100 \times \{1 - \exp[-(k_1/u)z_0]\}. \tag{5}$$

This justifies the empirical relation, Eq. (3), and hence the v_3 band absorption could be used to measure the degree of silane dissociation. Further, best fitting of the result of Fig. 12 to Eq. (5) leads to an estimation of the mean decomposition cross section of SiH_4 $<\sigma> \simeq 10^{-17}$ cm^2 and the mean dissociation threshold energy $\bar{\varepsilon} = 8.9$ eV. The spatial distribution of un-

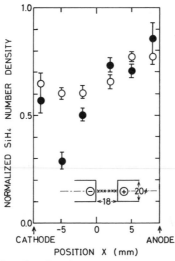

FIG. 13. Spatial distribution of undissociated silane in a dc glow discharge at pressures of 0.5 (●) and 1.0 (O) Torr. (From Hata *et al.*, 1983.)

dissociated silane $N(z)$ in the direction of gas flow is obtained from Eq. (4) and Fig. 12. Such a spatial distribution of undissociated silane exists also in the direction perpendicular to the electrodes in a flat-bed reactor, being directly determined by coherent anti-Stokes–Raman spectroscopy (CASR) (Fig. 13) (Hata *et al.,* 1983). The result indicates that the decomposition reactions in the dc discharge proceed more frequently near the cathode.

III. Reactors

5. CLASSIFICATION OF REACTORS

The glow-discharge deposition of a-Si : H has extensively been performed using various types of plasma reactors. Most of the systems currently used could be classified as tube reactors with external electrodes and flat-bed reactors with internal electrodes. The inductively coupled tube reactor has often been distinguished from the capacitively coupled system, but this may not be an appropriate classification, because even in the case of the so-called inductively coupled system the rf power is coupled capacitively through the dielectric wall, as explained later (Vossen, 1979).

a. Tube Reactors

A schematic of the two different techniques for coupling rf power into a tube reactor is shown in Fig. 14. In the tube-inductive system the rf power is

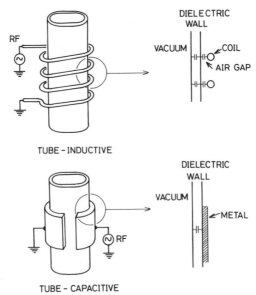

FIG. 14. Simplified sketch of tube reactors and their cross sections.

fed through nonuniformly distributed capacitors composed of air gap and dielectric wall. A plasma generated in a tube reactor is always separated by the dielectric wall from the rf electrodes, and hence ion bombardment on the internal wall will sometimes contaminate the discharge. Special care to avoid such contamination has to be taken by applying a magnetic field (Hamasaki *et al.*, 1980) or by employing a large tube diameter with respect to the dimension of the substrate (Vossen, 1979). The spatial inhomogeneities of electric field in the tube system are not significant when the operating pressure is in the range of a few Torr. The substrate holder is either grounded or floating.

b. Flat-Bed Reactors

The substrates are generally placed on a grounded table, mainly because the powered rf electrode is self-biased and suffers ion bombardment, which deteriorates the film quality. In this system a series capacitor is usually connected between the powered electrode and the rf generator, as shown in the inset of Fig. 1, and the self-bias is developed by charging the series capacitor. As described in Section 1, the grounded substrates are more or less bombarded by ions because of the presence of the plasma potential (Fig. 2). For minimizing such bombardment, the internal powered electrode should be small in comparison with the area of all other surfaces in contact with the discharge. Furthermore, the interelectrode spacing should be at least five times as large as the thickness of the ion sheath to avoid obstruction of the discharge, and the applied rf power should be as low as possible when the highest possible pressure is used. In the case of a dc glow discharge system, the substrates are usually placed on the powered electrode (cathode) because the deposition occurs mainly on the cathode surface. To avoid the positive ion bombardment damage, a cathode screen is added near the substrate (see Uchida, Chapter 3, this volume).

c. Magnetron Configuration

Motion of electrons in the plasma is primarily determined by the spatial distribution of electric field and easily influenced by magnetic field. The magnetic field parallel to the electric field reduces the transverse diffusion of electrons and results in the confinement of the positive column, while the magnetic field perpendicular to the electric field induces the cycloidal motion of electrons, which remarkably increases the ionization rate of gas molecules. As shown in Fig. 15a, the electron trajectory between the parallel plate electrodes can be calculated for the magnetic field in the direction of the electrode plane. The electron path is given by

$$x = mE_a/eB_z^2(\omega_0 t - \sin \omega_0 t), \tag{6a}$$

$$y = mE_a/eB_z^2(1 - \cos \omega_0 t). \tag{6b}$$

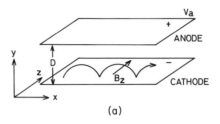

(a)

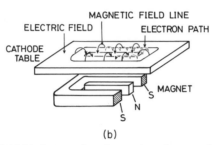

(b)

FIG. 15. (a) The cycloidal trajectory of an electron in a plane-parallel electrode magnetron. (b) Permanent magnets are attached below cathode table to produce in-plane magnetic flux.

Here B_z is the magnetic field strength, m is the free electron mass, e is the electron charge, E_a is the electric field strength V_a/D, and ω_0 is the angular cyclotron frequency eB_z/m. The magnitudes of the cathode voltage and the magnetic field determine whether the electrons move around the cathode in approximate cycloidal paths or whether they reach the anode. To realize cycloidal paths, application of a magnetic field as high as a few hundred gauss is sufficient for most cases of the glow discharge. The planar magnetron-type reactor is therefore constructed as Fig. 15b, where permanent magnets are attached to the rear surface of cathode. This type of cathode has first applied to an rf-sputtering apparatus (Thornton and Penfold, 1978) and then to a reactive ion etching system (Horiike *et al.*, 1981). Such a magnetron-type deposition system may be useful to increase the dissociation rate of reactive gas and to reduce the ion bombardment on the grounded table.

IV. Representative Deposition Techniques by rf Glow Discharge†

The influence of deposition parameters on the structure and electronic properties of rf plasma-deposited a-Si : H has been systematically studied by

† For a discussion of dc glow discharge see Chapter 3 by Y. Uchida.

the Xerox group (Street *et al.,* 1978; Knights, 1979). They measured the variation of the luminescence intensity, spin density, hydrogen content, and vibrational spectrum when the deposition parameters—rf power, gas pressure, gas concentration, bias voltage, and substrate temperature—are changed in a wide range. The role of electric and magnetic bias in rf plasma has been reported by several groups (Knights, 1979; Tsai and Fritzsche, 1979; Okamoto *et al.,* 1980; Taniguchi *et al.,* 1980; Martins *et al.,* 1981). It is also reported that hydrogen bond configurations in a-Si:H are remarkably influenced by gas flow rate (Hirose *et al.,* 1981).

6. DEPOSITION SYSTEM

Various types of flat-bed reactors are most widely used in the preparation of a-Si:H mainly because the reaction chamber can be easily scaled up without any substantial change in the basic construction of the system. Figure 16 shows a schematic of a capacitively coupled reactor. Here the powered electrode K is water cooled, while the grounded table A is heated up to ~300°C. The rf discharge is sustained between the electrodes K and A, whose spacing is mostly 4–5 cm, depending on the operating pressure. Silane gas is admitted to the reaction chamber with a typical flow rate of 20–200 sccm. For the purpose of doping, either phosphine or diborane are premixed with silane and the dopant concentration incorporated in the matrix is approximately equal to the molar fraction of doping gas to silane,

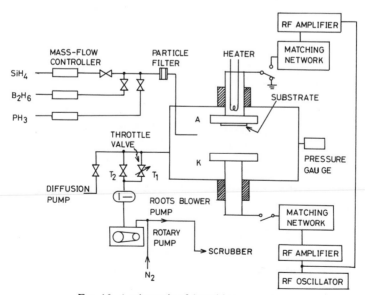

FIG. 16. A schematic of deposition apparatus.

TABLE III

Typical Range of Discharge Parameters

Parameter	Range
SiH₄ concentration	10–100% in H₂ or Ar
Total gas flow rate	20–200 sccm
Total pressure	0.05–2.0 Torr
rf power	1–100 W
Substrate temperature	200–300°C
Substrate bias	Mostly zero

although doped atoms do not necessarily occupy the substitutional sites (Spear and LeComber, 1976). Evacuation of the reactor before deposition is done either by a diffusion pump with a liquid nitrogen trap or by a roots blower pump, and then the system pressure for deposition is settled at an appropriate value by tuning both the gas flow rate and the pumping rate through the roots blower pump. The pumping speed is controlled by the use of the throttle valve T_1 or the stop valve T_2. When the substrate table A, which is usually grounded, is connected with an rf power source, a negative dc bias develops on the substrate surface during growth (Street *et al.*, 1978).

A typical range of deposition parameters employed in an ordinary apparatus is summarized in Table III.

7. Deposition Variables and Material Properties

A suitable set of deposition parameters has to be chosen in order to obtain well-defined and high-quality a-Si:H films. For the purpose of material characterization, the luminescence intensity, spin density, refractive index, infrared absorption spectrum, and so on, have been measured as functions of rf power, silane concentration, bias voltage, and flow rate.

The relation between the luminescence intensity and rf power is shown in Fig. 17, together with the luminescence intensity versus silane concentration curves (Street *et al.*, 1978). With the exception of 100% silane, the intensity decreases with the same logarithmic dependence on power and increases proportionally to the square root of silane concentration. Figure 18 shows plots of spin density (dangling bonds) against rf power and silane concentration. The spin density is definitely large at high power and low silane concentration. Comparing Fig. 17 with Fig. 18, there is a clear correlation between the presence of defect centers giving rise to spin and the existence of a competing nonradiative transition that lowers the luminescence intensity. It is likely that a-Si:H films produced in the silane depletion regime, which refers to a high-power and low-silane-concentration region, possess many

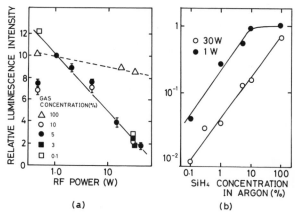

FIG. 17. Luminescence intensity in a-Si : H, $T_s = 230°C$ (a) as a function of rf power; (b) as a function of SiH_4 concentration. Both (a) and (b) at $10°K$. (From Street et al., 1978.)

more defect states than those obtained in silane excess regime. The bias effect on structural properties is shown in Figs. 19 and 20 (Knights, 1979). Note that a negative bias voltage appearing on the substrate is approximately proportional to power being coupled through the substrate table (-40 V at a power of 30 W). The density deficit is reflected in the refractive index variation shown in Fig. 19 and dense films can be prepared at low rf power. The density deficit is remarkably improved for biased samples consistent with Fig. 18a. It is interesting to compare Fig. 19 with Fig. 20, in

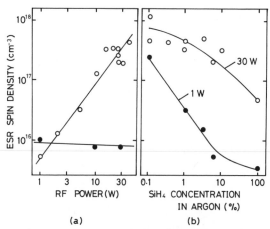

FIG. 18. Electron spin density in a-Si : H (a) as a function of rf power (●, biased; ○, unbiased. $g = 2.0055$); (b) as a function of silane concentration. Biased samples were produced by applying rf power on the substrate table (see Fig. 16). $T = 300°K$. (From Knights, 1979.)

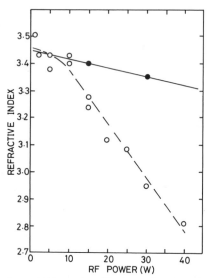

FIG. 19. Refractive index at 1500 nm (in a-Si : H) as a function of rf power ($T_s = 230°C$, 5% SiH_4/Ar). O, unbiased; ●, biased. (From Street *et al.*, 1978.)

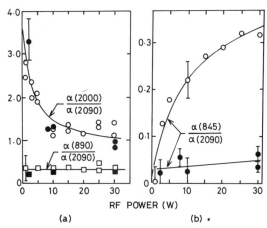

(a)   (b)

FIG. 20. Normalized absorption constants for Si–H vibrational modes: (a) Si–H stretch ($2000 cm^{-1}$) and SiH_2 scissors ($890 cm^{-1}$) normalized to SiH_2 stretch ($2090 cm^{-1}$). ●,■: biased; O, □; unbiased. (b) $(Si–H_2)_n$ wagging mode ($845 cm^{-1}$) to SiH_2 stretch ($2090 cm^{-1}$). O, unbiased; ●, biased. (From Knights, 1979.)

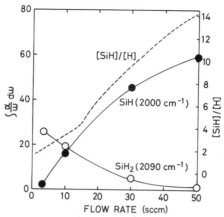

FIG. 21. Integrated absorption of Si–H and Si–H$_2$ stretching modes as a function of total flow rate. The dashed lines refers to the concentration ratio of the emissive SiH to H [SiH]/[H]. rf power = 20 W. $P = 0.64$ Torr. (From Hirose *et al.*, 1981.)

which the Si–H infrared absorption coefficient α is normalized to the Si–H$_2$ stretching mode absorption. The increase in the wagging mode absorption at 845 cm^{-1} could be associated with the decrease in the refractive index and the increase in defect density. The gas flow rate is also an important parameter that influences the hydrogen bond configurations, as shown in

TABLE IV

PHYSICAL PROPERTIES OF a-Si:H

Material constant	Typical value
Dark conductivity	$3 \times 10^{-9} \ \Omega^{-1} \ cm^{-1}$
Conductivity activation energy	0.76 eV
Photoconductivity (AM-1, 100 mW cm^{-2})	$1 \times 10^{-3} \ \Omega^{-1} \ cm^{-1}$
Optical band gap	1.7–1.8 eV
Temperature coefficient of optical gap	2.7×10^{-4} eV °K^{-1}
Electron mobility	0.5–1.0 cm^2 V^{-1} sec^{-1}
Hole mobility	$(1 \times 10^{-3} – 5 \times 10^{-3})$ cm^2 V^{-1} sec^{-1}
Carrier diffusion length	$> 1.0 \ \mu m$
Electron affinity	3.93 eV
Refractive index	3.43
Density	2.2 g cm^{-3}
Spin density	$< 1 \times 10^{16}$ spin cm^{-3}
Hydrogen content	18 at. %
Crystallization temperature	675°C

Fig. 21. Predominant incorporation of monohydride bonds and elimination of dihydride bonds can be achieved in high-flow-rate regimes. According to the optical emission study, this is explained in terms of the chemical reactions that proceed on the growing surface (Hirose *et al.*, 1981). Typical material constants of discharge produced a-Si: H are given in Table IV. These values can be obtained with good reproducibility at the present state of art for plasma deposition.

8. CROSS ELECTRIC FIELD CONFIGURATION

The refractive index and hydrogen bonding of a-Si: H are appreciably changed with a dc bias superimposed in the direction of rf electric field, as shown in Figs. 19 and 20. In the cross field plasma deposition system (Fig. 22), rf power is fed through the electrodes A and B and an additional dc bias is applied perpendicularly to the rf electric field using the electrodes C and D. Both the dc bias $V_b = V_D - V_C$ being varied from -150 to $+150$ V and the plasma potential V_p should determine the degree of the ion bombardment on the substrate surface. As a result, the material properties are dependent on the dc bias V_b, as shown in Figs. 23 and 24 (Hotta *et al.*, 1982). The negative substrate potential with respect to the plasma results in the predominant incorporation of monohydride bonds and a corresponding high photoconductivity is achieved. The detailed mechanism of the substrate potential effect in the rf plasma is not definitely known as yet.

9. MAGNETIC BIAS

The magnetic field in the direction parallel to the rf electric field confines the positive column and lowers the electron temperature in the plasma. This

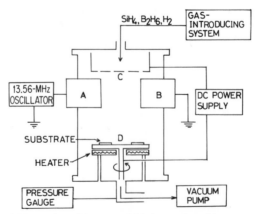

FIG. 22. Schematic illustration of the cross-field plasma deposition system. (From Okamoto *et al.*, 1980.)

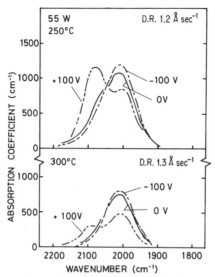

FIG. 23. IR absorption spectra at Si–H vibrational mode frequencies of undoped a-Si:H films with parameters of substrate temperature and bias voltage. (From Okamoto *et al.*, 1980.)

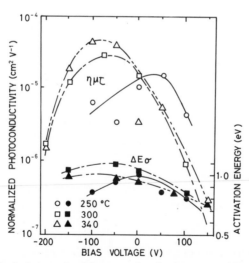

FIG. 24. Normalized photoconductivity $\eta\mu\tau$ and activation energy ΔE_σ of undoped a-Si:H versus bias voltage with a parameter of substrate temperature. rf power = 55 W. (From Okamoto *et al.*, 1980.)

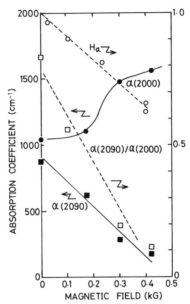

FIG. 25. Infrared absorption strength due to the Si–H (2000 cm^{-1}) and Si–H$_2$ (2090 cm^{-1}) bonds in a-Si:H prepared under magnetic field together with the optical emission intensity of the H$_\alpha$ (656 nm) line. (From Taniguchi, unpublished, 1980.)

situation is realized in the tube reactor as shown in Fig. 4b. The properties of a-Si:H deposited in the system of Fig. 4b are sensitively dependent on magnetic field applied. The Si–H stretching mode absorption due to the dihydride bonds (2090 cm^{-1}) decreases with increase in magnetic field, while the absorption due to the monohydride bonds (2000 cm^{-1}) increases (Fig. 25). It is worth noting that the optical emission intensity of the H$_\alpha$ (656 nm) line decreases as magnetic field increases, indicating the lowering of electron temperature. Along with appreciable decrease in the infrared absorption at 2090 cm^{-1} and increase of the 2000 cm^{-1} absorption, a pronounced increase in photoconductivity from 10^{-6} to 10^{-4} Ω^{-1} cm^{-1} at a flux of 10^{15} photons (cm^2 sec)$^{-1}$ takes place (Taniguchi *et al.*, 1980). A possible model to interpret this result is that the surface chemical reaction such as SiH + H → SiH$_2$ is suppressed as a consequence of decrease in the concentrations of both H and SiH radicals with the increase of magnetic field. The magnetic field parallel to the surface of the substrate table in the capacitively coupled rf system (see Fig. 15) also exhibits remarkable effect on dark conductivity and photoconductivity (Martins *et al.*, 1981).

10. DEPOSITION TECHNIQUE ASPECTS

Amorphous hydrogenated silicon is a structure-sensitive material as described in the previous sections, and therefore very fine tuning of deposition

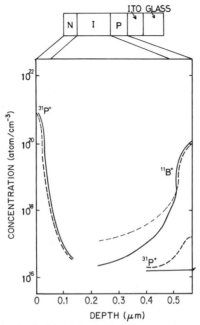

FIG. 26. IMA (ion microanalysis) depth profile of impurity concentration in p-i-n a-Si diodes. Solid lines and broken lines represent a-Si diodes prepared by consecutive, separated reaction chamber method, and by single reaction chamber method, respectively. (From Kuwano *et al.*, 1982.)

parameters and careful design of both reaction chamber and gas control system are needed for obtaining satisfactory reproducibility of the properties. The important points to grow well-defined a-Si:H will be as follows:

(1) Minimize contaminations from reactor walls (impurity gas desorption) and pumping system (oil back) as well.

(2) Keep the purity of silane that is often deteriorated by impurity gas desorption (mainly oxygen) from the internal tube wall of gas lines and by low-level air leakage in the gas control system. The leakage level should be maintained at less than 1×10^{-5} Torr liter sec.$^{-1}$

(3) Eliminate any spurious discharge between the powered electrode and its electric shield.

(4) Avoid the influence of autodoping from residual dopant atoms previously deposited on the reactor wall.

(5) Employ an appropriate in-process monitoring technique to identify the state of plasma.

In many cases of device applications, it is necessary to successively deposit undoped and doped layers. The autodoping effect in a single reaction chamber is significant when p, i, and n layers are deposited in this order as shown in Fig. 26. A consecutive, separated reaction chamber method can minimize such autodoping because each of doped and undoped layers is grown in the respective chamber separated by gate valves (Kuwano *et al.*, 1982).

11. A NEW CLASS OF SILICON-BASED AMORPHOUS MATERIALS

Flexible control of material properties is one of attractive research subjects in materials science, and amorphous hydrogenated silicon has good advantages in this respect because various ternally alloy systems such as a-Si_xC_{1-x}:H, a-Si_xO_{1-x}:H, a-Si_xGe_{1-x}:H and a-Si_xN_{1-x}:H can be synthesized by glow discharges. Another interesting system is microcrystalline hydrogenated silicon. These silicon-based materials can offer new possibilities of producing wide varieties of heterojunction and graded-gap structures.

Study of an additive to a-Si:H was first reported by Anderson and Spear (1976). They prepared a-Si_xC_{1-x}:H by the glow discharge of a $SiH_4 + C_2H_4$ gas mixture and showed that the optical band gap reaches a maximum at a compositional ratio of $x = 0.32$. They also found that the optical gap of a-Si_xN_{1-x}:H deposited from the $SiH_4 + NH_3$ plasma decreases from 5 to 2.5 eV as the molar fraction N_{NH_3}/N_{SiH_4} decreases from 20 to 5. The resulting films, however, were no longer semiconducting. Investigation on discharge-produced a-Si_xO_{1-x}:H showed that the optical absorption edge and luminescence peak energy shift toward higher photon energies as oxygen content increases (Knights *et al.*, 1980).

Remarkable improvements of the electronic properties of a-Si_xC_{1-x}:H and the substitutional doping have been achieved using the plasma deposition of a $SiH_4 + CH_4$ gas mixture instead of the $SiH_4 + C_2H_4$ system (Tawada *et al.*, 1982). As shown in Fig. 27, the AM-1 photoconductivity σ_{ph} for undoped specimens rapidly decreases as the optical band gap increases, while dramatic improvement of σ_{ph} occurs by boron doping. Such photoconductivity recovery is similarly seen in phosphorus-doped a-Si_xC_{1-x}:H, which is accompanied by a decrease in the spin density. Boron-doped a-Si_xC_{1-x}:H is successfully utilized as a window material of pin junction solar cells (Tawada *et al.*, 1981).

More recently, semiconducting a-Si_xN_{1-x}:H has been prepared by adding NH_3 to the $SiH_4 + H_2$ plasma. It is found that the optical bandgap is controllable over the range 1.75–5.5 eV by changing the composition x, and high photoconductivity is obtained at $x \simeq 0.8$ (Kurata *et al.*, 1981). The substitutional doping of this system is basically possible (Fig. 28) (Kurata *et al.*, 1982). However, the doping efficiency for phosphorus atoms has to be improved. It should be noted that undoped a-Si_xN_{1-x}:H is thermally stable up to 400°C and exhibits no Staebler–Wronski effect when $x \simeq 0.8$.

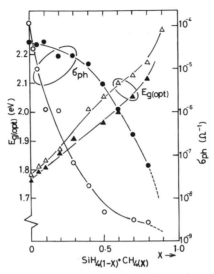

FIG. 27. Photoconductivity σ_{ph} and optical band gap $E_{g(opt)}$ of undoped ($\triangle$, $\bigcirc$) and boron-doped ($\blacktriangle$, $\bullet$) a-SiC:H films prepared by the decomposition of $[SiH_{4(1-x)} + CH_{4(x)}]$ gas mixture. (From Tawada *et al.*, 1982.)

The highest conductivity achievable in doped a-Si:H is at most $\sim 10^{-2}$ Ω^{-1} cm^{-1} and further increase of the maximum conductivity appears to be difficult for presently available deposition technique. Usui and Kikuchi (1979) have produced microcrystalline (μc)-like Si:H films by the glow discharge of $SiH_4 + PH_3 + Ar$ and obtained the very high conductivity of 25 Ω^{-1} cm^{-1} at a doping ratio of 2.5×10^{-2}. Extensive work on plasma-deposited μc-Si:H revealed that the necessary conditions to obtain the microcrystalline phase are to employ relatively high rf power and low concentrations of silane (Matsuda *et al.*, 1980; Hamasaki *et al.*, 1980). The optical band gap of doped μc-Si:H is almost identical to that of undoped a-Si:H and the highest conductivity exceeds 27 Ω^{-1} cm^{-1} for phosphorus doping and 7.8 Ω^{-1} cm^{-1} for boron doping (Hamasaki *et al.*, 1980, 1981a) (see Fig. 29). Optical emission studies on the growth of μc-Si:H indicated that the nucleation threshold of microcrystallites is correlated with the optical emission intensity ratio of SiH (414 nm) to H (656 nm) (Mishima *et al.*, 1982). In other words, excited hydrogen radicals that attack the growing surface might remove weak Si–Si bonds and scavenge hydrogen bonds on the surface, resulting in nucleation of the crystalline silicon network. High doping efficiency in μc-Si:H can be achieved also by the dc glow discharge of $SiH_4 + H_2$ (Y. Uchida, private communication, 1981). The other experiment to obtain μc-Si:H has been attempted for the first time by employing the chemical transport of silicon in low-pressure hydrogen

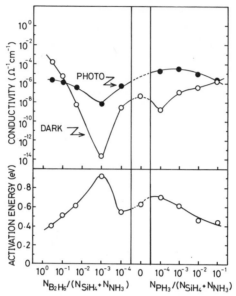

FIG. 28. Conductivity and its activation energy, and photoconductivity of a-Si$_x$N$_{1-x}$:H plotted against doping ratio. $N_{NH_3}/N_{SiH_4} = 0.26$. (From Kurata *et al.*, 1982.)

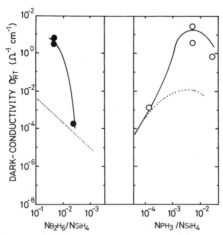

FIG. 29. Conductivity versus doping ratio for μc-Si:H (solid curves) and for a-Si:H (dashed curves after Spear and LeComber.).

plasma (Vepřek and Mareček, 1968). The structure of plasma-deposited μc-Si : H is considered to be a mixed phase, being composed of microcrystallites and connecting tissue of amorphous hydrogenated silicon (Mishima *et al.*, 1982). The volume fraction of this material is less than 100%, whereas that of μc-Si : H deposited by the chemical transport appears to be near 100% (Vepřek *et al.*, 1981).

V. Future Problems

In the early stage of fundamental research on physical properties of a-Si : H, the growth rate as low as ~ 1 Å sec^{-1} being typical for the silane glow discharge appears to be enough to obtain the necessary thickness of the thin films. However, recent progress in the device applications such as low-cost photovoltaic cells and electrophotography requires dramatic increase of the deposition rate in order to achieve the high throughput. For instance, at least 15-μm-thick films are necessary for conventional a-Si : H photoreceptors. This implies that the deposition rate should be more than 30 Å sec^{-1} from the practical point of view. Almost the same deposition rate will be needed even for photovoltaic applications. It will be a challenging research subject to realize extremely high growth rate without any deterioration of the material quality such as that previously shown in Table IV. In this regard, fundamental understanding of chemical reactions in the silane (or higher silane) plasma will provide helpful information about how to improve the deposition rate and how to control the material properties.

The other important problem unsolved is the light soak degradation (Staebler–Wronski effect) that occurs in bulk a-Si : H and in the device structures as well. A correlation between photocreation defects and incorporated or penetrated oxygen contaminants has been suggested (Carlson, 1982; Fritzsche, 1982), but the detailed mechanism of defect creation is not known as yet. Anyway, very careful study has to be carried out to clarify what kinds of extrinsic impurities and/or local structures introduced in a-Si : H during deposition are responsible for such structural and/or electronic instabilities.

In this chapter there is no detailed discussion of apparatus such as tunnel reactors with a perforated shield to eliminate substrate bombardment (Vossen, 1979), magnetron plasma reactors, and microwave discharge systems including ECR (electron cyclotron resonance) reactors (Matsuo and Kiuchi, 1983), because there is very little work on these machines despite the potential importance as new reactors under development. The present author did not provide too many details on currently available plasma reactors and the related experimental results, because the plasma deposition technique is still in "chaos" in many respects. Indeed there are a lot of controversial discussions on the deposition chemistry as well as on the characterization of materials. If you find out something interesting regarding the deposition techniques and new silicon-based materials described in this chapter, it would be better to refer to the original papers for understanding what happens in the plasma and what characteristics are expected for the material newly synthesized.

In conclusion, numerous variants of plasma reactors are available to prepare a-Si:H, but at the present state the optimization of deposition parameters is still necessary for a given reactor although fundamental knowledge on the discharge phenomena and currently reported results can suggest the general direction to develop the deposition technique. The substrate in contact with the plasma should more or less suffer the bombardment by ions and high-energy particles, and there is no definite answer as to what energy and flux densities of bombardment particles, if they are important, are favorable for preparing high-quality a-Si:H. Extensive studies on plasma chemistry are most important for the further development of preparation techniques.

REFERENCES

Anderson, D. A., and Spear, W. E. (1976). *Philos. Mag.* [8] **35,** 1.
Bourquard, S., Erni, D., and Mayor, J. M. (1981). *In* "Proceedings of the Fifth International Symposium on Plasma Chemistry" (B. Waldie and G. A. Farnell, eds.), p. 664. Heriot-Watt University, Edinburgh, Scotland.
Carlson, D. E. (1982). *Proc. U.S.-Jpn. Jt. Semin. Technol. Appl. Tetrahedral Amorphous Solids, 1982.*
Chen, F. F. (1965). *In* "Plasma Diagnostic Techniques" (R. H. Huddlestone and S. L. Leonard, eds.). Academic Press, New York.
Dang-Nue, M., Pierre, G., and Saint-Loup, R. (1974). *Mol. Phys.* **28,** 447.
Fritzsche, H. (1982). *Proc. U.S.-Jpn. Jt. Semin. Technol. Appl. Tetrahedral Amorphous Solids, 1982.*
Hamasaki, T., Kurata, H., Hirose, M., and Osaka, Y. (1980). *Appl. Phys. Lett.* **37,** 1084.
Hamasaki, T., Kurata, H., Hirose, M., and Osaka, Y. (1981a). *Jpn. J. Appl. Phys.* **20,** L84.
Hamasaki, T., Hirose, M., Kurata, H., Taniguchi, M., and Osaka, Y. (1981b). *Jpn. J. Appl. Phys., Suppl.* **20-1,** 281.
Hamasaki, T., Hirose, M., and Osaka, Y. (1981c). *J. Phys., Suppl.* **10,** C4-807.
Hamasaki, T., Hirose, M., and Osaka, Y. (1982). *Proc. Symp. Ion Sources Ion-Assisted Technol. 6th, 1982* p. 263.
Hata, N., Matsuda, A., and Tanaka, K. (1983). *Jpn. J. Appl. Phys.* **22,** L1.
Hayashi, T., Miyamura, M., and Komiya, S. (1982). *Jpn. J. Appl. Phys.* **21,** L755.
Hirose, M. (1982a). *Jpn. J. Appl. Phys., Suppl.* **21-1,** 275.
Hirose, M. (1982b). *Proc. U.S.-Jpn. Jt. Semin. Technol. Appl. Tetrahedral Amorphous Solids, 1982.*
Hirose, M., Hamasaki, T., Mishima, Y., Kurata, H., and Osaka, Y. (1981). *In* "Tetrahedrally Bonded Amorphous Semiconductors" (R. A. Street, D. K. Biegelsen, and J. C. Knights, eds.), p. 10. Am. Inst. Phys., New York.

Horiike, Y., Okano, H., Yamazaki, T., and Horie, H. (1981). *Jpn. J. Appl. Phys.* **20**, L817.
Hotta, S., Nishimoto, N., Tawada, Y., Okamoto, H., and Hamakawa, Y. (1982). *Jpn. J. Appl. Phys., Suppl.* **21-1**, 289.
Kampas, F. J., and Griffith, R. W. (1980). *Sol. Cells* **2**, 385.
Kampas, F. J., and Griffith, R. W. (1981). *J. Appl. Phys.* **52**, 1285.
Knights, J. C. (1979). *Jpn. J. Appl. Phys., Suppl.* **18-1**, 101.
Knights, J. C., Street, R. A., and Lucovsky, G. (1980). *J. Non-Cryst. Solids* **35–36**, 279.
Knights, J. C., Schmitt, J. P., Perrin, J., and Guelachvili, G. (1982). *J. Chem. Phys.* **76**, 3414.
Kocian, P. (1980). *J. Non-Cryst. Solids* **35–36**, 195.
Kurata, H., Hirose, M., and Osaka, Y. (1981). *Jpn. J. Appl. Phys.* **20**, L811.
Kurata, H., Miyamoto, H., Hirose, M., and Osaka, Y. (1982). *Jpn. J. Appl. Phys., Suppl.* **21-2**, 205.
Kuwano, Y., Ohnishi, M., Tsuda, M., Nakashima, Y., and Nakamura, N. (1982). *Jpn. J. Appl. Phys.* **21**, 413.
Martins, R., Dias, A. G., and Guimaraes, L. (1981). *In* "Tetrahedrally Bonded Amorphous Semiconductors" (R. A. Street, D. K. Biegelsen, and J. C. Knights, eds.), p. 36. Am. Inst. Phys., New York.
Matsuda, A., Nakagawa, K., Tanaka, K., Matsumura, M., Yamasaki, S., Okushi, H., and Iizima, S. (1980). *J. Non-Cryst. Solids* **35–36**, 183.
Matsuo, S., and Kiuchi, M. (1983). *Jpn. J. Appl. Phys.* **22**, L210.
Milligan, D. E., and Jacox, M. E. (1970) *J. Chem. Phys.* **52**, 2594.
Mishima, Y., Miyazaki, S., Hirose, M., and Osaka, Y. (1982). *Philos. Mag.* [*Part*] B **46**, 1.
Okamoto, H., Yamaguchi, T., Nitta, Y., and Hamakawa, Y. (1980). *J. Non-Cryst. Solids* **35–36**, 201.
Spear, W. E., and LeComber, P. G. (1976). *Philos. Mag.* **33**, 935.
Street, R. A., Knights, J. C., and Biegelsen, D. K. (1978). *Phys. Rev. B* **18**, 1880.
Taniguchi, M., Hirose, M., and Osaka, Y. (1980). *J. Non-Cryst. Solids* **35–36**, 189.
Tawada, Y., Kondo, M., Okamoto, H., and Hamakawa, Y. (1981). *Conf. Rec. IEEE Photovoltaic Spec. Conf.* **15**, 245.
Tawada, Y., Tsuge, K., Kondo, M., Okamoto, H., and Hamakawa, Y. (1982). *J. Appl. Phys.* **53**, 5273.
Thornton, J. A., and Penfold, A. S. (1978). *In* "Thin Film Process" (J. L. Vossen and W. Kern, eds.), p. 76. Academic Press, New York.
Tsai, C. C., and Fritzsche, H. (1979). *Sol. Energy Mater.* **1**, 29.
Turban, G., Catherine, Y., and Grolleau, B. (1980). *Thin Solid Films* **67**, 309.
Turban, G., Catherine, Y., and Grolleau, B. (1982). *Plasma Chem. Plasma Process.* **2**, 61.
Usui, S., and Kikuchi, M. (1979). *J. Non-Cryst. Solids* **34**, 1.
Vepřek, S., and Mareček, V. (1968). *Solid-State Electron.* **11**, 683.
Vepřek, S., Iqbal, Z., Oswald, H. R., Sarott, F. A., and Wagner, J. J. (1981). *J. Phys., Suppl.* **10**, C4-251.
Vossen, J. L. (1979). *J. Electrochem. Soc.* **126**, 319.

SEMICONDUCTORS AND SEMIMETALS, VOL. 21A

CHAPTER 3

dc Glow Discharge

Yoshiyuki Uchida

SEMICONDUCTOR LABORATORIES
FUJI ELECTRIC
CORPORATE RESEARCH AND DEVELOPMENT, LTD.
NAGASAKA, YOKOSUKA CITY
JAPAN

I. Introduction

In recent years, glow-discharge techniques have found increasing application in the fabrication of semiconductor devices (e.g., photoresist removal, plasma etching, and deposition of passivating films). In addition to these, an important application has been developed in the synthesis of the semiconductor itself, an amorphous silicon – hydrogen alloy, currently referred to as hydrogenated amorphous silicon (a-Si: H), which we now know contains typically between 5 and 40 at. % hydrogen. Discharge-produced a-Si: H has remarkably good semiconductor qualities in comparison with unhydrogenated a-Si prepared by conventional sputtering or evaporation. This is due to the low density of dangling bonds and the correspondingly low density of localized states in the gap (Spear and LeComber, 1972).

The first attempt at producing a-Si: H was performed by an inductive radiofrequency (rf) glow-discharge system (Chittick *et al.*, 1969). This system utilized an external coil surrounding a cylindrical glass chamber in

41

which silane (SiH_4) was decomposed by rf power fed from the coil and a-Si:H films were deposited on substrates. However, difficulty in scale-up may be a major problem with this system. Another type of rf system has been used for depositing a-Si:H films, in the reaction chamber SiH_4 is decomposed by discharge between two electrodes using capacitively coupled rf power (Knights, 1976). Because of its ability to deposit uniform films over large areas, this type of system is most widely used at present and has been applied to the mass production of a-Si:H solar cells for consumer electronics — as in hand-held calculators (Uchida *et al.*, 1981a; Kuwano and Ohnishi, 1981).

Good-quality a-Si:H films may also be obtained by direct-current (dc) glow-discharge decomposition of SiH_4. The arrangement inside the reaction chamber is basically similar to that of the rf capacitive system, but instead of the rf generator, a dc power supply is connected to the electrodes. The first a-Si:H films used to make electronic devices were produced by dc glow discharge in SiH_4 (Carlson, 1977) and a-Si:H solar cells with an area of 49 cm², demonstrated as the first large-area cells, were also made in a dc glow-discharge system (Ichimura *et al.*, 1979). Several types of a-Si:H solar cells with excellent photovoltaic performance have been produced in a dc glow discharge (Carlson, 1981).

This review will concentrate on the apparatus and techniques of dc glow discharge for depositing amorphous silicon films including a-Si:H and a-Si:F,H, and the correlation between deposition conditions and properties of the resultant films based on published data. The distinct advantages of this method and the important problems for further development of techniques will also be discussed.

II. General Description

1. APPARATUS AND PROCEDURE

A dc glow-discharge system is shown schematically in Fig. 1a. Figure 1b shows a schematic of the proximity dc glow-discharge system that will be described later. The system is composed of a gas supply system, a reaction chamber, a vacuum system and a dc power supply. In the reaction chamber an anode and a cathode of metal plate or screen are placed facing each other not more than 10 cm apart. Since the a-Si:H film is deposited mainly onto substrates placed on the cathode (in Fig. 1a), we will refer to this technique as cathodic dc glow discharge as distinct from proximity dc glow discharge. The cathode contains a heater to heat the substrates to a temperature of 100–400°C in order to obtain films suitable for electronic devices as discussed later. SiH_4 gas, which is sometimes diluted with H_2 or Ar gas, is fed

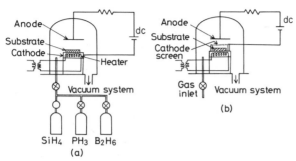

FIG. 1. Schematics of a-Si:H deposition systems: (a) cathodic dc glow discharge, (b) proximity dc glow discharge.

into the reaction chamber, with the vacuum system operating to maintain a constant gas pressure and remove waste gases. The discharge is powered by a dc electric field. In the discharge, hot electrons accelerated by the field interact with the SiH_4 and a variety of new species are generated: ions, neutral atoms, and molecules in both ground and excited states. These are transported toward the substrate surface to form a-Si:H. The deposition mechanism for discharge-produced a-Si:H is not definitely known as yet. However, current models will be presented in other sections of this book.

2. PROXIMITY DC GLOW DISCHARGE

The dc electric field accelerates positive ions in the SiH_4 discharge toward the cathode and causes positive ion bombardment of the cathodic substrates. This tends to produce additional defects in the a-Si:H film (Carlson et al., 1980). The proximity dc glow discharge was proposed as a means of eliminating ion bombardment damage (Carlson, 1977). In this method, the SiH_4 discharge is produced between the anode and a cathode screen located ~1–2 cm above the substrate, as shown in Fig. 1b, and no electric field exists in the vicinity of the substrate. If necessary, the substrate can be biased with respect to the potential of the cathode screen; this facilitates the control of ionic species reaching the substrate. Another advantage of the proximity dc glow-discharge method is a relatively fast deposition rate of a-Si:H film onto the insulating substrate compared to that in the cathodic dc glow-discharge method, which is very slow.

3. DEPOSITION CONDITIONS

The dc potential necessary to maintain stable discharge between anode and cathode, referred to here as self-sustaining voltage, depends on gas

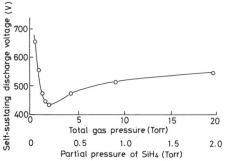

FIG. 2. Self-sustaining discharge voltage as a function of total gas pressure ($SiH_4/H_2 = \frac{1}{10}$) and partial pressure of SiH_4. (From Uchida *et al.*, 1981a.)

pressure and anode-to-cathode spacing as well as on the kinds of gas and electrode material used. Figure 2 shows the dependence of self-sustaining voltage on total gas pressure (10% SiH_4 in H_2) and partial pressure of SiH_4 (Uchida *et al.*, 1981a). The electrodes were stainless steel plates and the gap was 4 cm. The self-sustaining voltage decreases sharply with gas pressure to a certain minimum value and increases gradually with increasing gas pressure. A gas pressure larger than that giving the minimum self-sustaining voltage is usually chosen in the deposition of a-Si:H films in order to keep the discharge stable against small fluctuations of the SiH_4 pressure.

The deposition rate of a-Si:H film on a stainless steel substrate in the cathodic glow discharge in 10% SiH_4 diluted with H_2 was systematically evaluated as a function of discharge current density, gas pressure, substrate

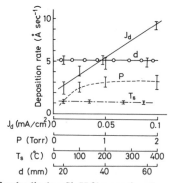

FIG. 3. Deposition rate of cathodic dc a-Si:H film as a function of discharge current density, J_d, ($P = 2$ Torr, $T_s = 270°C$, and $d = 40$ mm); partial pressure of SiH_4, P, ($J_d = 0.01$ mA cm^{-2}, $T_s = 270°C$, and $d = 40$ mm); substrate temperature T_s, ($J_d = 0.01$ mA cm^{-2}, $P = 0.2$ Torr, and $d = 40$ mm), and anode to cathode spacing, d, ($J_d = 0.05$ mA cm^{-2}, $P = 2$ Torr, and $T_s = 270°C$).

TABLE I

TYPICAL RANGES OF PARAMETERS FOR DC
GLOW-DISCHARGE DEPOSITION OF a-Si:H

Parameter	Range
Substrate temperature	$100-400°C$
SiH_4 gas pressure	$0.1-2$ Torr
Discharge current	$0.01-2$ mA cm^{-2}
$SiH_4/(SiH_4 + H_2)$	$0.1-1$

temperature, and anode-to-cathode spacing (Ichimura *et al.,* 1980). In the experiment all parameters except one variable were constant. The deposition rate varies with both discharge current density and gas pressure, as shown in Fig. 3. Although the maximum deposition rate shown in Fig. 3 is about 10 Å sec^{-1} when the discharge current density is 0.1 mA cm^{-2} and the partial pressure of SiH_4 is 2 Torr, it increases proportionally to the discharge current: The film is deposited at about 100 Å sec^{-1} with a 1 mA cm^{-2} discharge current density. This high deposition rate is an advantage of the cathodic dc glow discharge. On the other hand, the deposition rate in the proximity dc glow discharge is roughly one order of magnitude less than that in the cathodic dc glow discharge.

Typical ranges of deposition parameters for the dc glow-discharge deposition of a-Si:H are summarized in Table I.

4. OTHER DC DISCHARGE TECHNIQUES

In addition to the cathodic and proximity dc glow-discharge methods, two other techniques utilizing dc discharge have been proposed for making a-Si:H films. These techniques will be described here in outline only because the detailed experimental data have not yet been reported.

One is the multipole dc discharge method (Drevillon *et al.,* 1980). In this method a SiH_4 plasma is sustained by fast electrons emitted from hot tungsten filaments in a partially confining multipolar magnetic structure, with a-Si:H films being deposited on the walls. At very low SiH_4 pressures ($0.1-5$ mTorr) the deposition rate remains in the range of $1-10$ Å sec^{-1}.

Another recently proposed method is the Penning glow discharge of SiH_4 (Hirao *et al.,* 1981). A glow discharge is generated between two cathodes surrounded by a cylindrical anode that is connected to the positive output voltage of a dc power supply. A magnetic field is applied perpendicularly to the substrate surface during the deposition of a-Si:H films to form a dense plasma. The substrate to be deposited can be attached to both cathodes. Since the Penning discharge enables the glow discharge to be maintained in a stable mode at low pressure compared to conventional methods, the plasma can be confined within the electrodes by lowering the pressure to 0.07 Torr, thus reducing the reaction chamber wall – plasma interaction.

III. Properties of Films Made by dc Glow Discharge

5. HYDROGEN IN a-Si:H

As mentioned in the introduction, the hydrogen incorporated in the a-Si:H film saturates dangling bonds and modifies the whole network. Therefore, both the concentration and incorporation scheme of hydrogen affect the film properties.

The major parameters in the dc glow discharge affecting hydrogen incorporation are the substrate temperature and the discharge power density (Zanzucchi et al., 1977; Ichimura et al., 1980; Carlson et al., 1982). In a-Si:H films deposited by any of the glow-discharge techniques described in the introduction, the hydrogen content decreases with increasing substrate temperature. Figure 4 shows a compositional profile of a layered sample deposited using a proximity dc glow discharge in a $SiH_4 - PH_3$ mixture (4% PH_3 in SiH_4) where the substrate temperature was varied in steps (Carlson et al., 1982). The hydrogen and phosphorus concentrations were determined by means of secondary ion mass spectrometry (SIMS). As shown in Fig. 4, the hydrogen content decreases in steps as the substrate temperature increases in steps. This trend is also observed in both undoped and boron-doped films.

Infrared absorption measurement is a good method for obtaining compositional information such as content and species of bonded hydrogen in a-Si:H films (Brodsky et al., 1977). Table II shows a tentative assignment of the various vibration modes in the infrared absorption spectra of a-Si:H

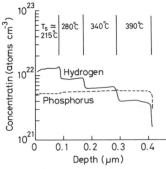

FIG. 4. A compositional profile of hydrogen and phosphorus in a layered sample where the substrate temperature was varied in steps. All for 90 sec. (From Carlson et al., 1982.)

TABLE II

VIBRATION MODES FOR THE INFRARED
ABSORPTION SPECTRA OF a-Si:H FILM[a,b]

Structural group	Bond stretching	Bond bending	Bond rocking
SiH	2000	—	630
SiH$_2$	2090	880	630
SiH$_3$	2140	860,905	630
(SiH$_2$)$_n$	2090 (2090–2100)	845,890	630

[a] From Fritzche (1980).
[b] In units of inverse centimeters.

films (Fritzsche, 1980).† The absorption peaks attributed to such groups as SiH$_2$, SiH$_3$, and (SiH$_2$)$_n$ increase as the substrate temperature decreases for both films made in a capacitive rf glow discharge (Knights, 1979) and in a dc glow discharge (Ichimura *et al.*, 1980). Films tend to be more polymerlike as deposition temperature is lowered, and they lose their good semiconductor qualities. On the other hand, too high a substrate temperature results in an increase of unsaturated dangling bonds in the resultant film (Knights, 1979). Therefore, a-Si:H films for electronic devices must be deposited on a substrate heated to a temperature between 100 and 400°C, as shown in Table I.

The discharge power density also influences the infrared absorption spectra of the resultant film, as shown in Fig. 5 (Ichimura *et al.*, 1980). Films were deposited on a crystalline silicon wafer at 270°C by cathodic glow discharge in SiH$_4$ at 2 Torr total pressure (10% SiH$_4$ in H$_2$). An increase of discharge power density results in the appearance of an absorption peak at about 880 cm^{-1} and shift of the stretching vibrational peak near 2000 cm^{-1} to a higher wavenumber. This is accompanied by photoconductivity reduction, which may be partly caused by an increase in positive ion bombardment. Generally speaking, the a-Si:H film deposited with lower discharge power exhibits better quality.

Other parameters such as gas pressure and gas flow rate also influence hydrogen incorporation, but the effects are generally smaller than that of the substrate temperature or discharge power density.

Out-diffusion of hydrogen from a-Si:H films is observed when the film is heated to temperatures greater than the deposition temperature for $T_s >$ 250°C, resulting in the deterioration of the film properties (Fritzsche, 1980).

† A discussion of infrared vibrational spectra by P. J. Zanzucchi will be found in Chapter 4 of Volume 21B of this treatise.

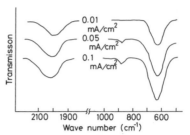

FIG. 5. Infrared transmission spectra of a-Si:H films and the dependence on discharge current density for film growth. $T_s = 270°C$. (From Ichimura *et al.*, 1980.)

The diffusion coefficient for deuterium in an a-Si:H film deposited in a cathodic glow discharge was reported as shown in Fig. 6 (Carlson and Magee, 1978). For predicting the operational life of a-Si:H devices, these data are quite useful because there should be little difference in diffusion coefficients between deuterium and hydrogen in a-Si:H. It was suggested that the out-diffusion of hydrogen is not significant for a-Si:H at 100°C until after more than 10^4 years (Carlson and Magee, 1978).

6. OPTICAL PROPERTIES

The conductivity of discharge-produced a-Si:H film at room temperature typically ranges from 10^{-12} to 10^{-8} $(\Omega\ cm)^{-1}$ in the dark and increases sharply when the film is illuminated. Figure 7 shows the absorption coefficient as a function of photon energy for films deposited in a cathodic glow discharge at different substrate temperatures (Zanzucchi *et al.*, 1977). The

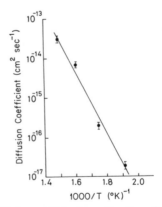

FIG. 6. The deuterium diffusion coefficient as a function of $1000/T$ for cathodic dc films. $E_a \simeq 1.53 \pm 0.15$ eV. (From Carlson and Magee, 1978.)

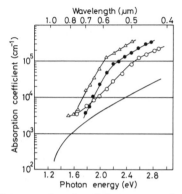

FIG. 7. Absorption coefficient as a function of photon energy for cathodic dc films prepared with substrate temperatures of 210 (○), 325 (●), and 415°C (△). Data are compared to crystalline Si. (From Zanzucchi *et al.*, 1977.)

absorption coefficient is much greater than that of crystalline Si in the visible wavelength region. This is the reason that an efficient photovoltaic device can be constructed utilizing a-Si : H of less than 1 μm in thickness. By using the absorption coefficient (α) data in Fig. 7, $(\alpha h\nu)^{1/2}$ is plotted as a function of photon energy ($h\nu$), as shown in Fig. 8 (Zanzucchi *et al.*, 1977). Data for films made in a inductive rf glow discharge are also shown in the figure. The values of the optical band gap obtained from the intersection of the lines decrease as the substrate temperature is raised. The changes are primarily due to the variation of hydrogen content in the films with the substrate temperature, as described in Part III, Section 5.

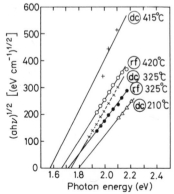

FIG. 8. Optical gap for a-Si : H produced at different substrate temperatures in a cathodic or inductive rf glow discharge. (From Zanzucchi *et al.*, 1977.)

7. DOPING OF a-Si:H

The electronic properties of discharge-produced a-Si:H can be controlled by a substitutional impurity doping in the gas phase (Spear and LeComber, 1975). The undoped material is slightly n-type, but by introducing phosphine (PH_3) or diborane (B_2H_6) into the reaction chamber during glow discharge of SiH_4, the Fermi level of the resultant film can be moved with respect to the band edges and the conductivity increases correspondingly. In other words, the material can be doped n or p-type. Substitutional doping is one of the most important factors in the application of a-Si:H to electronic devices. The conductivity of doped film depends not only on the concentration of dopant in the gas phase but also on such parameters as substrate temperature and discharge power density (Ichimura et $al.,$ 1980; Carlson, 1981). A high substrate temperature and low discharge power density tend to increase the conductivity of the resultant film. When the substrate temperature is varied from 210 to 390°C in a proximity dc glow discharge using B_2H_6—SiH_4 mixture (1.3% B_2H_6), the conductivity of the boron-doped films increases from 10^{-4} to 10^{-2} (Ω cm)$^{-1}$ (Carlson et $al.,$ 1982).

IV. Preparation and Properties of a-Si:F,H

Amorphous silicon containing both fluorine and hydrogen (a-Si:F,H) has been suggested as material for thin-film photovoltaic devices (Ovshinsky and Madan, 1978). Since then the effect of fluorine incorporated in amorphous silicon on the photovoltaic performance has been examined by many researchers (Konagai and Takahashi, 1980; Madan et $al.,$ 1980; Carlson and Smith, 1981) but has not yet been sufficiently clarified.

The structure and properties of a-Si:F, H film also depend on the deposi-

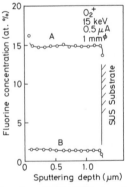

FIG. 9. SIMS profiles of fluorine for a-Si:F,H films produced by (A) cathodic dc and (B) proximity dc glow discharge. (From Uchida et $al.,$ 1982.)

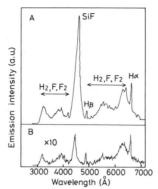

FIG. 10. Optical emission spectra from (A) cathodic dc and (B) proximity dc glow discharge of a SiF_4-H_2 misture. (From Uchida *et al.*, 1982.)

tion method. Figure 9 shows SIMS profiles of fluorine in films deposited with cathodic dc and proximity dc glow discharges in a SiF_4-H_2 mixture $(SiF_4/H_2 = 7)$ (Uchida *et al.*, 1982). The films were deposited on a substrate at a temperature of 300°C with a discharge current density of 0.13 mA cm^{-2}. The fluorine content of the films varies greatly according to the deposition method and is 15 at. % and 1.5 at. % for the cathodic dc and proximity dc films, respectively, while the hydrogen content varies less (9-15 at. %). Figure 10 shows the optical emission spectra (OES) from the cathodic dc and proximity dc glow discharges within 1 cm of the substrate. Emissions from such excited species as SiF, H_2, H, F_2, and F are commonly observed in both OES, but the emission intensity from the proximity dc glow discharge is considerably smaller than that of the cathodic dc glow

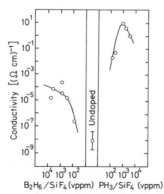

FIG. 11. Conductivity as a function of $B_2H_6-SiH_4$ or PH_3-SiH_4 for a-Si:F,H films produced by cathodic dc glow discharge. (From Nishihata *et al.*, 1981.)

discharge. The smaller amount of excited species in the proximity dc glow discharge is due to the absence of electric field in close vicinity to the substrate, as discussed in Part II, Section 2. The difference in SiF, F_2, and F content between the two kinds of glow discharge reflects the difference in fluorine content of the resultant films.

Figure 11 shows conductivity plotted as a function of doping level of PH_3 and B_2H_6 for n-type and p-type a-Si : F,H films, respectively (Nishihata et al., 1981). The films were deposited in a cathodic dc glow discharge in a $SiF_4 - H_2$ mixture with a substrate temperature of 300°C. The conductivity of the n-type film increases more sharply with increasing dopant content of the gas than that of the p-type film. This is due to microcrystallization of the film (Matsuda et al., 1980), which occurs both in a-Si : H and a-Si : F,H as a result of the high power density discharge (Hamasaki et al., 1980; Uchida et al., 1981b).

V. Concluding Remarks

The preceding description of the preparation of amorphous silicon by dc glow-discharge techniques has focused on deposition techniques and the relationship between deposition conditions and the composition and properties of the resultant materials.

One of the major advantages of the techniques described is the ability to form films suitable for photovoltaic devices, especially when the proximity dc glow discharge is adopted. As described previously, positive ion bombardment of films can be effectively minimized by locating a cathode screen in front of film substrates, and a-Si : H solar cells having good photovoltaic performance have been produced in proximity dc glow discharge systems (Carlson, 1981). Moreover, it has been reported quite recently that the carrier diffusion length of proximity dc a-Si : H films is 0.5 – 0.7 μm and is twice as large as that of capacitive rf films (McMahon, 1982).

It is necessary for a rf glow-discharge system to utilize a matching network that transforms the impedance of the rf generator to that of the load. Proper matching of impedance is necessary to obtain an efficient coupling of power to the discharge. On the other hand, only a dc power supply with a relatively low capacity is necessary for producing the discharge in a dc glow-discharge system. Moreover, no shielding technique for eliminating radiation interference is necessary in this kind of system. This simple setup is another advantage of dc glow discharge.

In order to apply the dc glow-discharge techniques to industrial production of amorphous silicon devices, the following important items require further development. At low discharge currents only part of the surface of electrodes is covered by a luminous glow in general; the surface is covered entirely as the current is raised. In a dc glow-discharge system, appropriate deposition conditions such as the power density, gas pressure, and anode-to-cathode spacing should be chosen in order to obtain a uniform discharge covering the electrodes entirely. Furthermore, anode-to-cathode spacing must be kept exactly constant over the whole area of the electrodes. No data have been published so far on the uniformity of a-Si : H film produced in a dc glow-discharge system that has a wide deposition area on an industrial scale (1 m^2 or more).

In a proximity dc glow-discharge system, a-Si : H film is also deposited on the cathode screen. This insulates the cathode screen and results in a reduction of discharge current. Moreover, the a-Si : H deposited on the cathode screen may flake off and can be a cause of pinholes in the films. This problem will occur during long runs of film deposition.

These technical problems will be overcome in the near future. Devices made of a-Si : H, with the exception of solar cells for consumer electronics, are still in the development stage. The great advantages of a-Si : H as a material for thin-film devices will be fully realized when the density of localized states in the gap can be controlled more precisely. In order to do this, efforts to understand in detail the deposition mechanism and defects of a-Si : H are essential.

REFERENCES

Brodsky, M. H., Cardona, M., and Cuomo, J. J. (1977). *Phys. Rev. B: Solid State* [3] **15**, 3556–3571.
Carlson, D. E. (1977). U. S. Patent 4,064,521.
Carlson, D. E. (1981). *Prog. Cryst. Growth Charact.* **4**, 173–193.
Carlson, D. E., and Magee, C. W. (1978). *Appl. Phys. Lett.* **33**, 81–83.
Carlson, D. E., and Smith, R. W. (1981). *Conf. Rec. IEEE Photovoltaic Spec. Conf.* **15**, 694–697.
Carlson, D. E., Magee, C. W., and Thomas, J. H., III (1980). *Sol. Cells* **1**, 371–379.
Carlson, D. E., Smith, R. W., Magee, C. W., and Zanzucchi, P. J. (1982). *Philos. Mag. [Part] B* **45**, 51–68.
Chittick, R. C., Alexander, J. H., and Sterling, H. F. (1969). *J. Electrochem. Soc.* **116**, 77–81.
Drevillon, B., Huc, J., Lloret, A., Perrin, J., deRosny, G., and Schmitt, J. P. M. (1980). *Appl. Phys. Lett.* **37**, 646–648.
Fritzsche, H. (1980). *Sol. Energy Mater.* **3**, 447–501.
Hamasaki, T., Kurata, K. Hirose, M., and Osaka, Y. (1980). *Appl. Phys. Lett.* **37**, 1084–1086.
Hirao, T., Mori, K., Kitagawa, M., Ishihara, S., Ohno, M., Nagata, S., and Watanabe, M. (1981). *J. Appl. Phys.* **52**, 7453–7455.
Ichimura, T., Kitamura, A., Ohira, K., Takeda, Y., and Saga, M. (1979). *Abstr. 26th Spring Meet Jpn. Soc. Appl. Phys. Relat. Soc.* p. 469 (in Japanese).
Ichimura, T., Uchida, Y., and Nabeta, O. (1980). *Ext. Abstr., Electrochem. Soc. Meet.* **80-2**, 1423–1425.
Knights, J. C. (1976). *Philos. Mag.* [8] **34**, 663–667.
Knights, J. C. (1979). *Jpn. J. Appl. Phys.* **18**, *Suppl.* **18-1**, 101–108.

Konagai, M., and Takahashi, K. (1980). *Appl. Phys. Lett.* **36**, 599–601.

Kuwano, Y., and Ohnishi, M. (1981). *J. Phys. (Orsay, Fr.)* **42**, *Suppl.* **10**, C4-1155–C4-1164.

Madan, A., McGill, J., Czubatyj, W., Yang, J., and Ovshinsky, S. R. (1980). *Appl. Phys. Lett.* **37**, 826–828.

McMahon, T., and Konenkemp, R. (1982). *Conf. Rec. IEEE Photovoltaic Spec. Conf.* **16**, 1389–1393.

Matsuda, A., Yamasaki, S., Nakagawa, K., Okushi, H., Tanaka, K., Iizima, S., Matsumura, M., and Yamamoto, H. (1980). *Jpn. J. Appl. Phys.* **19**, L305–L308.

Nishihata, K., Komori, K., Konagai, M., and Takahashi, K. (1981). *Jpn. J. Appl. Phys.* **20**, *Suppl.* **20-2**, 151–155.

Ovshinsky, S. R., and Madan, A. (1978). *Nature (London)* **276**, 482–483.

Spear, W. E., and LeComber, P. G. (1972). *J. Non-Cryst. Solids* **8-10**, 727–738.

Spear, W. E., and LeComber, P. G. (1975). *Solid State Commun.* **17**, 1193–1196.

Uchida, Y., Sakai, H., Ichimura, T., and Haruki, H. (1981a). *Fuji Electr. Rev.* **27**, 34–40.

Uchida, Y., Ichimura, T., Ueno, M., and Ohsawa, M. (1981b). *J. Phys. (Orsay, Fr.)* **42**, *Suppl.* **10**, C4-265–C4-268.

Uchida, Y., Ichimura, T., Nabeta, O., Takeda, Y., and Haruki, H. (1982). *Jpn. J. Appl. Phys.* **20**, *Suppl.* **20-2**, 193–198.

Zanzucchi, P. J., Wronski, C. R., and Carlson, D. E. (1977). *J. Appl. Phys.* **48**, 5227–5236.

CHAPTER 4

Sputtering

T. D. Moustakas

CORPORATE RESEARCH LABORATORIES
EXXON RESEARCH AND ENGINEERING COMPANY
ANNANDALE, NEW JERSEY

I. Introduction

There are two plasma methods for depositing hydrogenated amorphous silicon: glow-discharge decomposition of silane and reactive sputtering from a silicon target in a mixture of inert gas and hydrogen. Of the two, glow discharge has attracted considerably more attention. The extensive investigation of this deposition process, through plasma diagnostic studies, and of the resulting films through structural, optical, transport, recombination, and device studies, has led to the fabrication by Catalano *et al.* (1982) of solar cell structures with efficiency up to 10%. It is widely believed that this rapid progress in the device area resulted from the identification of thin-film growth conditions, which led to material with the lowest structural and compositional inhomogeneities.

The alternative technology of reactive sputtering, inherently, has many advantages as a deposition process. Principal among these is the ability to uncouple the two source materials, Si and H, and independently to optimize

55

the role that each plays in forming the network and in determining the ensuing electro-optical properties. This type of control is not possible for glow-discharge decomposition of silane because the source materials are initially chemically bonded in the feed gas SiH_4. In addition, sputtered films in general are mechanically strong (good film – substrate adhesion) and the process is easily scalable. Despite these obvious advantages, this process has not been pursued as vigorously as the glow-discharge process, primarily because early work suggested that sputtered a-Si : H films are inferior in their electronic properties to those produced by the glow-discharge process. In addition, there is a widespread belief that sputtering is an unsuitable technology for surface-sensitive devices.

In the past two years significant progress has been made in the understanding of the thin-film growth process of the sputtered material, and this was followed by progress in the device area. Solar cell structures with efficiency up to 5.5% have recently been announced (Moustakas, 1983a,b).

This chapter reviews progress made over the past few years in the preparation of a-Si:H films by the method of rf diode reactive sputtering. The material is organized to include a brief historical background of the evolution of this method and sections on physical and chemical effects in sputter deposition of a-Si:H films, on the mechanisms of film growth, and on methods to control structural and compositional inhomogeneities. Finally, we comment on the device potential of this method and on future research directions.

II. Historical Background

The development of amorphous silicon by sputtering can be divided historically into two eras. Prior to 1974 sputtering was widely used for the deposition of pure tetrahedrally coordinated amorphous semiconductors. Structural studies on such materials by Moss and Graczyk (1969), Shevchik and Paul (1972), and Temkin et al. (1973) have revealed that such films contain microvoids on the order of 5 – 10 Å in diameter. Optical, transport, and electron spin resonance measurements have been plausibly and self-consistently interpreted by Brodsky et al. (1970) and Paul et al. (1973) on a model that assumed that these structural defects give rise to a large density of states in the gap of the semiconductor. At the same time, the studies of Spear (1974) indicated that a-Si produced by glow-discharge decomposition of SiH_4 has a considerably smaller density of states in the gap and exhibited electronic properties typical of semiconducting behavior. It has been speculated by Brodsky et al. (1970) that the difference between sputtered and glow discharge a-Si might be caused by hydrogen contamination.

Against this background Lewis et al. (1974) have produced the first sputtered hydrogenated amorphous germanium films by sputtering from a germanium target in a mixture of argon and hydrogen. This initial study and subsequent studies of such material by Lewis (1976), Connell and Pawlik (1976), Moustakas and Connell (1976), and Moustakas and Paul (1977) indicated that hydrogen incorporation reduced the density of spins by orders of magnitude with corresponding changes in the optical, transport, and recombination properties.

Hydrogen incorporation was found to have far more dramatic effects in amorphous silicon than in amorphous germanium. Following the discovery by Spear and LeComber (1975) that glow discharge material could be doped n-type and p-type, Paul et al. (1976) have demonstrated that sputtered hydrogenated amorphous silicon could also be doped n-type and p-type and that semiconductor junctions could be formed. Subsequent studies of this material by Moustakas et al. (1977), Anderson et al. (1977, 1979), and Moustakas (1979) suggested that careful optimization of the deposition parameter space could lead to material electronic properties similar to that produced by glow-discharge decomposition of silane.

Following these initial discoveries, the focus shifted to understanding the bonding configuration of hydrogen in the amorphous silicon network and to identify those deposition parameters that control hydrogen bonding. Experimental tools used in these studies include Si–H vibrational studies, photoemission, hydrogen evolution, and nuclear magnetic resonance. Such studies have recently been reviewed by Paul and Anderson (1981).

The vibrational studies of Brodsky et al. (1977) and Freeman and Paul (1978) identified the stretching modes in the $2000–2100$-cm^{-1} region, the bending modes in the $800–900$-cm^{-1} region, and the rocking and wagging modes at 630 cm^{-1}. These studies indicated that in sputtered material, produced under a variety of deposition conditions, the stretching vibration is always a doublet. It is generally accepted that the mode at 2000 cm^{-1} is due to isolated SiH units. The assignment of the mode at 2100 cm^{-1} is still controversial. Brodsky et al. (1977) suggested that it is due to SiH$_2$ units. Paul (1980) and Shanks et al. (1980) have proposed that this mode could also arise from SiH units in particular local environments. Jeffrey et al. (1979, 1980) demonstrated that sputtered a-Si:H films having only the 2000-cm^{-1} stretching mode can be made by sputtering at high power with the substrates located below the edge of the dark space. Oguz et al. (1981) also reported that films grown on substrates located next to the dark space have the Si–H stretching vibration at 2000 cm^{-1}. Moustakas (1982) developed bias sputtering as a means of controlling the bonding of hydrogen. This method will be discussed in greater detail in Part VI.

Photoemission studies on sputtered a-Si:H were first reported by von Roedern et al. (1979). These studies revealed the existence of a number of peaks in the valence-band density of states which were correlated with SiH, SiH_2, and SiH_3 configurations.

Information regarding the bonding of H in the silicon network can also be obtained by hydrogen evolution studies. Such studies on sputtered a-Si : H films were performed by Oguz et al. (1980; Oguz and Paesler, 1980). These authors find that samples produced at low hydrogen partial pressure show a single evolution peak near 600°C, whereas those produced at high partial pressure of hydrogen have also a second peak near 375°C. Although these data suggest the existence of two sites for hydrogen attachment, their interpretation is not firm yet. For more details the reader is referred to works by Oguz (1981) and Bruyere et al. (1980).

Nuclear magnetic resonance measurements in sputtered a-Si:H films were reported by Jeffrey et al. (1981) and Carlos et al. (1981). These studies revealed two superposed resonance lines of different widths as was also observed for glow discharge a-Si : H by Reimer et al. (1980). The narrower line contains about 2–4 at. % H and was attributed to monohydride configurations. The width of this line suggests that some of the monohydride units are clustered rather than randomly distributed. The broader line contains the rest of the hydrogen and is associated with densely clustered monohydrides or polyhydrides. The interpretation of these data and their correlation with the results obtained from other hydrogen probing experimental techniques is not firm yet.

In parallel with these studies, a large volume of work on optical transport and recombination properties has been generated in a number of laboratories. These properties were found to be sensitive to the hydrogen content and to the microstructure of the films. A comprehensive review on these properties was published by Paul and Anderson (1981).

Methods have been developed to probe the states in the gap of the semiconductor. Such methods include ESR, field effect, capacitance – voltage, optical absorption, and drift mobility. These studies indicated that device-quality material, having density of states in the middle of the gap of 10^{15} or lower, can be fabricated by the method of sputtering (Tiedje et al., 1981).

Thin-film growth and microstructural studies have been reported by Anderson et al. (1979), Moustakas (1979, 1982), Jeffrey et al. (1979), Tiedje et al. (1981), Moustakas et al. (1981a), Martin and Pawlewicz (1981), Turner et al. (1981), and Ross and Messier (1981a,b). In these studies the kinetics of the film growth and the influence of physical and chemical sputtering effects in controlling structural and compositional inhomogeneities have been investigated. These studies will be discussed in greater detail in the rest of the text.

Photovoltaic studies in this material were first reported by Thompson et al. (1978), Victorovitch et al. (1979), Moustakas et al. (1980), and Anderson *et al.* (1980). These studies were performed on Schottky barrier and metal–insulator–semiconductor devices. The moderate performance of such devices (conversion efficiency ~2%) has reinforced the widespread belief that sputtered a-Si:H films are inferior to those produced by glow-discharge decomposition of silane. However, recent progress in identifying deposition parameters that control film growth and progress in doping and device studies (Tiedje *et al.*, 1981; Moustakas *et al.*, 1981a,b, 1982, 1983; Morel and Moustakas, 1981; Moustakas and Friedman, 1982; Moustakas, 1982; Moustakas and Maruska, 1983; Maruska *et al.*, 1983) have led to the fabrication of solar cell structures with efficiencies up to 5.5% (Moustakas, 1983a,b).

III. Physical and Chemical Effects in Sputter Deposition of a-Si:H Films

In this section we review physical and chemical effects that are likely to influence the growth mechanism of a-Si:H films produced by rf diode sputter deposition. We are going to explore phenomena occurring at the target, in the glow discharge, and at the substrate. For a general reference on sputtering the reader is referred to the book by Vossen and Kern (1978).

1. PHYSICAL AND CHEMICAL EFFECTS AT THE TARGET

There are three types of phenomena that probably occur simultaneously at the Si target surface during sputtering in a mixture of Ar and H_2 (Sections 1a–1c).

a. Bombardment by Ar^+ and H^+ Ions

During this process a number of effects may occur.

(1) Some of the incident ions are neutralized and reflected as high-energy neutrals. The probability for this elastic backscattering depends on the ratio of M_I/M_T, where M_I is the mass of the impinging ion and M_T is the mass of the target atom. Ross and Messier (1981a) have noticed that momentum transfer considerations favor backscattering from the Si target of H but not of Ar atoms.

(2) The ion bombardment may cause the target to eject a secondary electron. These electrons are accelerated away from the target with an initial energy equal to the target potential.

(3) The ion may be implanted into the target. Argon implantation into Si is substantial even at low ion energies (Comas and Carosella, 1968).

(4) The target also emits UV and x-ray radiation.

(5) Ion bombardment may damage and possibly disorder the surface of the target.

(6) Ion bombardment removes neutral Si atoms by momentum transfer. This is the sputtering process that leads to the formation of the film. The threshold for sputtering single-crystal silicon is 10–15 eV (Wolski and Zdanuk, 1961). The sputtering yield by H^+ ions is probably very small.

(7) A significant fraction of the energy of the bombarding ions is dissipated in the target in the form of heat (Maissel, 1970).

b. Si: H Compound Formation

This process is favorable at high partial pressures of hydrogen (H_2/Ar $\gg$ 1) and low sputtering rates. It is well known in reactive sputtering (Schwartz, 1964) that under these conditions compound formation (SiH_x) occurs at the surface of the target as a solid state reaction. Later, some of this SiH_x is sputtered off. As discussed in Part IV such a process results in a significant decrease in deposition rate and in high secondary electron emission.

c. Chemical Sputtering

H-plasma etches crystalline Si (Vepřek, 1980). Therefore the hydrogen plasma may react with the surface of the target and form volatile SiH_x compounds. As discussed in Part IV the etching rate for such a process depends on the temperature of the Si target and on the concentration of atomic hydrogen. In the presence of such a process, the film grows by a combination of physical sputtering and glow-discharge decomposition of silane.

2. EFFECTS IN THE GLOW DISCHARGE THAT INFLUENCE FILM GROWTH

The density of the gas in the discharge controls the thermalization of the different particles (electrons, ions, neutrals) that are removed from the target. Depending on whether bombardment of the substrate by such particles is desirable or undesirable, one may choose to work at low or high gas pressure.

In an rf discharge, neutral sputtered atoms can be positively ionized by the Penning process (Coburn and Key, 1972). In this process metastable discharge atoms (Ar^m) transfer their energy and ionize neutral sputtered species (X) according to the reaction

$$Ar^m + X \rightarrow X^+ + Ar + e.$$

Silicon atoms will be readily ionized by such a process because their ionization potential (7 eV) is significantly smaller than the energy of metastable Ar atoms (11.55 and 11.72 eV). Such positively ionized silicon atoms will provide another source for bombardment of the substrate.

3. PHYSICAL AND CHEMICAL EFFECTS AT THE SUBSTRATE

The effects occurring at the substrate can be classified into the following categories.

a. Vapor Condensation and Network Formation

During this stage Si vapor from the target arrives at the substrate and condenses from an excited state. The atoms utilize their heat of condensation or energy supplied by other means to move around to final low energy sites. This motion is influenced also by the binding energy of the atom to the substrate. Indeed, different results are obtained with different substrates. In the process of hopping from one adsorption site to another, the atom may reevaporate or join with another atom to form a pair. This atomic pair has lower mobility. Therefore it is likely to be joined by other atoms, and the nucleation process of film growth has begun. This process leads initially to isolated islands, which as they grow in size coalesce to form a continuous film.

Movchan and Demshishin (1969) showed that the critical parameter in vacuum deposition is the ratio of the substrate temperature to the melting point of the evaporant (T_s/T_M). If this ratio is less than 0.45, the growth mode is columnar. Thornton (1974) has extended this model to the sputtering processes and included the role played by the sputtering gas. In this process a decrease in gas pressure is equivalent to an increase in substrate temperature and vice versa. Such effects will be discussed in detail in Part V.

b. Substrate Bombardment by Energetic Particles and Radiation

In addition to sputtered neutral species from which the film is grown, the substrate is also bombarded by numerous other species arriving from the target or extracted from the glow discharge. For example, secondary electrons, and conceivably secondary negative ions, are accelerated away from the target and by bombarding the substrate may cause substrate heating and radiation damage. Many of these species may be thermalized by collisions within the gaseous ambient, but a substantial number arrive at the substrate with full target potential. Also, since the substrate in a glow discharge is negatively charged with respect to the plasma, it is also bombarded by positive species extracted from the discharge. Such species include Ar^+ and H^+ ions as well as Si atoms ionized positively by the Penning process.

In the case of rf glow discharge these bombardment effects by charged particles are also influenced by the system's geometry. It is known (Vossen and Cuomo, 1978) that the ratio of the average potential on the target (V_T) to that of the grounded electrode (V_G) with respect to the plasma is given by

$$V_T/V_G = (A_G/A_T)^4, \tag{1}$$

where A_G and A_T are the areas of the grounded electrode and of the target, respectively. These areas refer to the areas in contact with the plasma. Therefore, the voltage of the substrate with respect to the plasma depends on the system geometry and can change for the same system by confining the discharge.

Bombardment effects by charged particles can also be influenced by an external bias intentionally applied to the substrate. The effects of both self-induced and externally supplied bias on the growth of a-Si : H films will be discussed in Parts V and VI.

In addition to bombardment by charged and neutral particles, the substrate is also subjected to UV and x-ray radiation from the target and glow discharge. These effects may influence the kinetics of the Si and H reactions at the surface of the substrate and can also induce radiation damage.

c. Si : H Compound Formation

As was discussed earlier, at high partial pressures of hydrogen ($H_2/A_r \gg 1$) and low sputtering rates, Si : H compound formation is likely to occur at the surface of the target. In the opposite regime, low hydrogen partial pressure ($H_2/A_r \ll 1$) and high sputtering rate, compound formation is likely to occur at the surface of the substrate. This process and the kinetics of the Si and H reactions are discussed in Part IV.

d. H-Plasma Etching of the Film

H plasma is known to etch amorphous silicon (Moustakas, 1982); therefore the effect of such a process is likely to influence the growth of the thin film. A discussion of this process is given in Part IV.

IV. Synthesis of Si : H Compounds

The mechanism of synthesis of Si : H compounds by sputtering from a Si target in an atmosphere of Ar and H_2 is poorly understood. In classical reactive sputtering (Schwartz, 1964; Vossen and Cuomo, 1978), reactions between the target material and the reactive gas occur either at the surface of the substrate or at the surface of the target, or both. Reactions in the gas phase are considered unfavorable because the simultaneous conservation of energy and momentum requires the presence of a third body. Whether the reaction occurs at the substrate or at the target depends on the partial pressure of the reactive gas and the target sputtering rate. At low reactive gas partial pressure and high target sputtering rate, virtually all the compound formation occurs at the substrate, whereas in the opposite regime, high reactive gas partial pressure and low target sputtering rate the compound formation occurs at the target. The transition from one regime to the other occurs at a certain threshold value of the reactive gas partial pressure; for metallic targets this is accompanied by a sharp decrease in the sputtering rate (Heller, 1973; Vossen and Cuomo, 1978). There are three factors that contribute to the decrease in the sputtering rate: (1) compounds, in general, have a lower sputtering yield than metals; (2) compounds have higher secondary electron emission yield than metals and therefore the energy of the incoming ions is used to produce and accelerate secondary electrons; (3) sputtering by reactive gas ions is less efficient than that by inert gas ions.

In addition to these reactive sputtering effects, the film growth may also be influenced by chemical sputtering effects. Hydrogen plasma is known to etch both crystalline silicon (Vepřek, 1980) and amorphous silicon (Moustakas, 1982). Since in rf diode sputtering both the surface of the target and the surface of the growing film are in contact with the plasma, the possibility that the hydrogen plasma attacks both and forms volatile SiH_x species is very likely. The etching rate of such processes was found to depend strongly on the density of atomic or ionic hydrogen and on the temperature of silicon. More specifically, the etching rate was found to attain a maximum value when the temperature of silicon is between 100 and 200°C and falls abruptly at lower and higher temperatures. This temperature dependence is probably related to the mobility of hydrogen atoms on the surface of silicon. These etching effects, particularly the etching of the growing film, may influence the film growth in a variety of ways. One possibility is that etching is anisotropic with higher rates at the intercolumnar low-density regions. Therefore such processes could affect the film microstructure. Another possibility is the modification of the film chemistry since volatile SiH_x species are likely to form more easily by interaction of atomic hydrogen with polyhydride configurations.

These effects have been investigated through spectroscopic studies. One study by Matsuda et al. (1980) reports the presence of SiH_4 in their sputtering plasma; another by Paesler et al. (1980) reports that SiH_4 is not present in their sputtering plasmas. In the following I discuss the results of these studies and suggest a possible resolution of the controversy.

Based on their observation, Matsuda and co-workers suggest the following mechanism for the thin-film growth: hydrogen gas introduced into the sputtering chamber is excited and decomposed into H atoms (or radicals) in the vicinity of the target where the electric field is the strongest. Such radicals react with sputtered Si atoms within the plasma close to the target, producing SiH_x molecules. After several collisions with other species, SiH_x molecules reach the surface of the substrate and are solidified into a-Si:H through surface reaction. The authors recognize that an alternative mechanism of forming SiH_x molecules is by direct etching of the Si target by H atoms, but they suggest that this must be a secondary mechanism, since the emission line intensity of SiH_4 is lowered in 100% H_2 gas.

Contrary to Matsuda and co-workers, I propose that the second mechanism is the most likely one for the formation of SiH_4 in their sputtering plasma because gas-phase reactions are not considered favorable. In such a model SiH_4 can be formed by H-plasma etching of the silicon target, or silicon deposited on the walls or substrate platform. As mentioned earlier, this etching mechanism depends strongly on the temperature of the silicon source. As an example, if the target is poorly water cooled, its surface may reach temperatures in excess of 100°C as a result of the sputtering process. It is estimated (Vossen and Cuomo, 1978) that 1% of the energy incident on the target surface goes into ejection of sputtered particles, 75% goes into heating of the target, and the remainder is dissipated by secondary electrons that bombard and heat the substrates. Under such circumstances the etching rate of the silicon target by the H plasma is expected to have its maximum value. The observation by the authors that the amount of SiH_4 decreases in 100% H_2 gas can also be accounted for in this model. The etching rate of silicon is proportional to the amount of atomic hydrogen, which has been shown by the authors to decrease in the absence of argon.

Paesler and co-workers, on the other hand, operated their sputtering apparatus with well-water-cooled target. As a result, the etching rate of Si by the H-plasma should be very small, and this should account for the lack of SiH_4 in their sputtering plasma.

On the basis of these observations, I propose the following model for film growth. If the target is well water cooled and the substrate platform is held either close to room temperature or above 200°C, there is very little H-plasma etching of either the target or the growing film. Under such experimental conditions, the kinetics of the Si and H reactions could be interpreted along the lines of classical reactive sputtering. If the target is held at temperatures between 100 and 200°C, then besides physical sputtering, the target is constantly etched by the H plasma and forms SiH_4. Under such conditions the film grows by a combination of physical sputtering and glow-discharge decomposition of SiH_4. If the target is water cooled and the substrates are held at temperatures between 100 and 200°C, the kinetics of the Si and H reactions are influenced not only by the physical sputtering processes but also by the etching of the growing film by the H-plasma.

To test this model we investigated the kinetics of the Si and H reactions (Tiedje *et al.,* 1981; Moustakas *et al.,* 1981a) under sputtering conditions that do not favor H-plasma etching of either the target or the growing film. (The target was water cooled and the substrate was held at temperatures higher than 200°C.) In these studies we deposited a number of films at low partial pressures of hydrogen ($H_2/Ar \ll 1$) and evaluated the Si and H reactions by measuring the density of states in the middle of the gap, which is related to Si dangling bonds and the hydrogen content in the films. Since these studies have been published we present only a brief summary of these results.

The samples were deposited at 275°C and argon pressure of 5 mTorr. The hydrogen pressure was varied from 0.3 to 1.4 mTorr. The density of states in the middle of the gap was measured by capacitance-voltage and ESR techniques (Tiedje *et al.*, 1981). The results are shown in Fig. 1 and clearly indicate that the density of states depends exponentially on the partial pressure of hydrogen in the discharge. The relative amount of bonded hydrogen in these films has been inferred from the integrated intensity under the Si–H stretching vibration (assumed to be proportional to the concentration of bonded hydrogen). The dependence of the integrated intensity of the Si–H stretching mode on the partial pressure of hydrogen is given in Fig. 2.

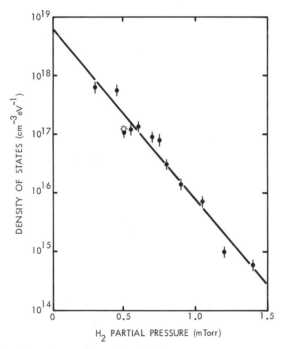

FIG. 1. Density of states in the middle of the gap as a function of hydrogen partial pressure. The error bars indicate the reproducibility of the measurements, and the solid line is a least-squares fit to the data. The point indicated by the symbol (☆) is the ESR spin density (1.1×10^{17} cm^{-3}). (After Tiedje *et al.*, 1981.)

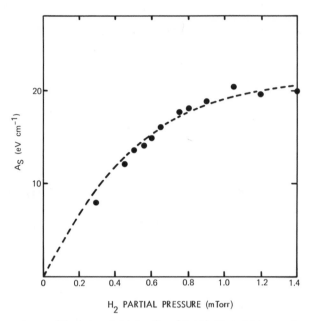

FIG. 2. Dependence of the integrated intensity of the Si–H stretching mode on the partial pressure of hydrogen. The dashed line is a fit to the data based on Eq. (4) as discussed in the text. (From Tiedje *et al.*, 1981.)

To model the Si and H reactions from these data, we postulated that the measured gap states are due to dangling silicon bonds. During the film growth process, a potential dangling bond at the growing surface of the films will either capture an H atom or be covered up by a layer of silicon atoms. Once the dangling bond is covered, it is unlikely to capture a hydrogen and becomes an electronically active defect. The capture rate of hydrogen atoms by dangling bonds will be given by the product $\sigma\theta F N$, of σ the capture cross section of the dangling bond for hydrogen, θ the sticking coefficient, F the hydrogen flux onto the surface of the film, and N the number of dangling Si bonds. For a given film growth rate R (monolayers per second), the dangling bond will have on the average a time of $1/R$ in which to capture an H atom before it is sealed up by an overlayer of silicon. The time dependence of the number of dangling bonds N on the growing surface of the film will be given by

$$dN/dt = -\sigma\theta F N. \qquad (2)$$

If N_0 is the number of dangling bonds present in the absence of hydrogen, then after the cutoff time $1/R$ the number surviving is given by

$$N = N_0 \exp(-\sigma\theta F/R). \tag{3}$$

Equation (3) predicts an exponential decrease in the density of dangling bonds with increasing H partial pressure since the H flux F is expected to be proportional to the partial pressure of H_2.

In order to fit the data of Fig. 1 quantitatively to Eq. (3), one needs to know which species of hydrogen (molecule, atom, ion) is the important one for hydrogen attachment to the dangling bonds. Such information is not available. However, the density of all these species is expected to scale with the H partial pressure, provided the argon pressure is high compared to the H pressure. For lack of more detailed information, we will use for F the flux of H_2 molecules in a hypothetical neutral gas at the same temperature and pressure. We find $F = A p_H$ where $A = 1.1 \times 10^{18}$ cm^{-2} sec^{-1} mTorr^{-1}. If we use 10^{-16} cm^2 for σ and the experimental value of R of about 1 monolayer per second, we find a sticking coefficient $\theta = 0.06$. This small value of sticking coefficient is reasonable if we assume that the atomic hydrogen is the active species and that the more abundant molecular hydrogen is not expected to eliminate dangling bonds.

Since the density of dangling bonds in the absence of hydrogen is much smaller than the actual hydrogen content of the hydrogenated material, it is clear that only a small fraction of the incorporated hydrogen eliminates potential dangling bonds. Most of the H bonds to other sites. We can estimate a sticking coefficient of these sites for H using the same kinetic model of thin-film growth process. In this model the hydrogen concentration will be given by

$$A = A_{max} [1 - \exp(-\sigma\theta' F/R)], \tag{4}$$

where A_{max} should reflect the maximum amount of H that can be incorporated into the lattice. By fitting the data in Fig. 2 with Eq. (4), we find that the H-sticking coefficient is $\theta' = 0.02$ if we use the same capture cross section (10^{16} cm^2) as for dangling-bond sites.

The fact that the sticking coefficient inferred from the H content is smaller than that inferred from the density of gap states may mean that most of the hydrogen has a high probability of being displaced by Si during the film growth process or that the H is incorporated into reconstructed dangling bonds. The sticking coefficient for these sites is likely to be smaller than that for true dangling bonds, because the hydrogen incorporation involves bond breaking. In this interpretation the bulk of the hydrogen is incorporated into sites that, in the absence of hydrogen, are reconstructed dangling bonds and do not have a spin or contribute states near midgap.

A further prediction of the model is that the hydrogen content in the films depends also on the deposition rate. To test this prediction we deposited three series of films at deposition temperature 275°C, argon pressure 15 mTorr, deposition rates between 1 and 4 Å sec^{-1} and studied their hydrogen content (Moustakas *et al.*, 1981a). The data of the hydrogen content in these films versus the ratio P_H/R are shown in Fig. 3. The dashed curve is the fit to the data based on Eq. (4) with a sticking coefficient 0.03. Therefore at these ranges of deposition rates the data are in good agreement with the model.

In the kinetic theory we have neglected thermal desorption of H from bonding sites and assumed that the H and Si are far from thermodynamic equilibrium. This assumption can be checked by comparing the H content as a function of P_H, for films deposited at different substrate temperatures with other parameters held fixed. The results for two series of films, one deposited at 385°C and one at 225°C, are shown in Fig. 4. The first thing to note is that the saturation H content at high P_H is higher for the low-temperature films. A plausible explanation of this result is that the growing Si network anneals more at 385°C than at 225°C. At the high temperature the Si network is likely to be less defective and, as a result, may have fewer H bonding sites.

The data in Fig. 4 suggest that the sticking coefficient θ at high temperature (385°C) is about equal to or larger than the sticking coefficient at 225°C. This observation rules out a thermodynamic equilibrium interpretation of the data. In this interpretation the effective sticking coefficient at a given P_H would be determined by the ratio of the H flux onto the surface to the rate of thermal desorption. The desorption process will involve bond breaking and hence should be thermally activated and strongly temperature dependent. In this case, the apparent sticking coefficient would decrease

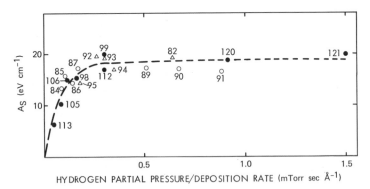

FIG. 3. Dependence of the integrated intensity of the Si–H stretching mode on the ratio of the partial pressure of hydrogen to the deposition rate. The dashed line is a fit to the data based on Eq. (4) as discussed in the text. $\triangle$, $R = 1.1$ Å sec^{-1}; $\bullet$, $R = 2.5$ Å sec^{-1}; $\bigcirc$, $R = 4.0$ Å sec^{-1}. (After Moustakas *et al.*, 1981a.)

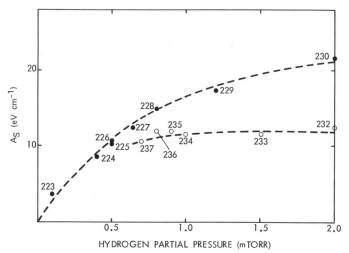

FIG. 4. Dependence of the integrated intensity of the Si–H stretching mode on the partial pressure of hydrogen for films produced at $T_s = 225°C$ (●) and $T_s = 385°C$ (○). (After Moustakas *et al.*, 1981a.)

dramatically from 225 to 385°C, contrary to the experimental results in Fig. 4, where the sticking coefficient appears to go in the opposite direction. Thus we conclude that the system is far from thermal equilibrium and the kinetic model is appropriate.

These studies demonstrated that the compound formation of films, produced at low partial pressure of hydrogen ($H_2/Ar \ll 1$) with substrate temperatures higher than 200°C and water-cooled target, occurs at the surface of the substrate. The stoichiometry of these films was found to depend on the relative rates of arrival at the substrate of Si and H atoms, in agreement with the predictions of the reactive sputtering theories.

It has been shown in another paper (Moustakas, 1982) that for $H_2/Ar \gg 1$) the Si:H compound formation occurs at the surface of the target. For details on these studies the reader is referred to the original work.

V. Evolution and Control of Structural Inhomogeneities

The structural studies of Knights and Lujan (1979) have led to the discovery that plasma deposition of a-Si:H films proceeds via nucleation and growth of island structures (average lateral dimensions ≃ 100 Å). The coalescence of these islands, if imperfect, leads to columnar morphology in subsequent growth. These authors report that growth of the films in pure silane, at low deposition rates, and deposition temperatures between 200 and 300°C leads to a more perfect coalescence of the islands. Films showing columnar morphology are unstable on exposure to the atmosphere. The interstitial regions between the columns can be penetrated by active atmospheric impurities leading to postdeposition contamination effects (Knights, 1979). The electronic properties of such films are poor, since the low-density interstitial regions are correlated with electronically active defects (Biegelsen et al., 1979).

Evidence of microstructural changes as a function of deposition parameters of sputtered amorphous silicon were first reported by Freeman and Paul (1978). These authors observed that films produced at high argon pressure show postdeposition contamination effects, which they consider evidence of film porosity.

Another study by Paulewicz (1978) on unhydrogenated amorphous silicon correlated defect creation to bombardment by Ar atoms, backscattered from the silicon target. This study concluded that the best-quality films (high resistivity) are those produced at high argon pressure (> 150 mTorr). Moustakas (1979) reinterpreted Paulewicz's result in terms of postoxidation contamination effects.

Anderson et al. (1979) studied in more detail the electronic properties of hydrogenated amorphous silicon films produced as a function of argon and reported that high-pressure films have poor electronic properties. These authors attributed the observed effects to the bombardment of the film by the neutral sputtered Si species from which the film grows. They argue that weak bombardment of the growing film might be beneficial in removing loosely bound material from the surface. At high argon pressures, this desirable bombardment is removed due to the faster collisional thermalization of the Si species.

The most comprehensive study on the microstructure of sputtered a-Si : H films has been reported by Ross and Messier (1981a,b). Their studies confirm that the growth habit of this material is also columnar, as in glow discharge decomposition of silane. Using TEM and SEM, they demonstrated that films produced at high argon pressures have columnar microstructure, while those produced at low argon pressure show no noticeable microstructure.

Ross and Messier (1981a,b) and Moustakas (1982) correlated these microstructural changes with corresponding changes in the self-bias voltage of the substrates and attributed these structural changes to charged particle bombardment effects.

In the rest of this section, I give some examples to illustrate the evolution and control of microstructure in sputtered or Si : H films (Moustakas, 1982). A number of films have been deposited at fixed hydrogen partial pressure ($P_H = 0.7$ mTorr) and argon pressure varying between 5 and 40 mTorr. The substrates were electrically floated and held at 275°C. The target was supplied with a constant power of 200 W. The choice of this mode of operation resulted in a small but monotonic increase of the deposition rate from 2.1 to 2.7 Å sec⁻¹, because the ion current increases as the pressure increases. As a result of that, the hydrogen content in the films was found to decrease with the argon pressure from 21.5 to 19.5 at. %. Microstructural changes in these films were inferred from measurements of the index of refraction and infrared vibrational spectra.

Figure 5 shows the dependence of the index of refraction n measured immediately after the film deposition at wavelength $\lambda \sim 2 \mu m$ versus the argon pressure in the discharge. The value of n remains constant up to about 20 mTorr of argon and then drops rather abruptly. The reduction of the index of refraction at high argon pressures is accompanied by postdeposition contamination. Figure 6 shows the infrared vibrational spectra for samples deposited at 5, 10, 20, and 40 mTorr of argon. The films deposited at 5 and 10 mTorr of argon show the usual Si–H vibrations, whereas the samples deposited at 20 and 40 mTorr of argon additionally show Si–O (~ 1100-cm⁻¹), Si–N (880-cm⁻¹), and C–H (~ 3000-cm⁻¹) vibrations. In agreement with Freeman and Paul (1978), we find that these additional vibrations are due to postdeposition contamination, and they grow as a function of time.

The reduction of the index of refraction for the films produced at high argon pressure can be accounted for as resulting from the film's porosity and from the formation of low index of refraction coatings (SiO_x, SiN_x) in the intercolumnar voids. Therefore, the results of Figs. 5 and 6 are consistent

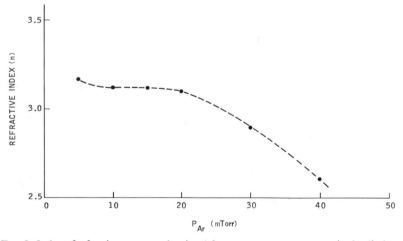

FIG. 5. Index of refraction, measured at $\lambda \simeq 1.9 \mu m$ versus argon pressure in the discharge. $T_s = 275°C$. $P_H = 0.7$ mTorr. (After Moustakas, 1982.)

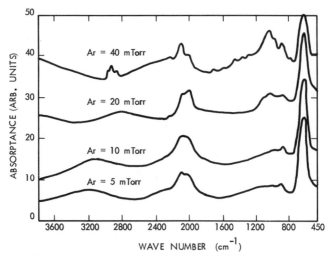

FIG. 6. Infrared vibrational spectra of samples deposited at argon pressure of 5, 10, 20, and 40 mTorr. (From Moustakas, 1982.)

with the observations of Ross and Messier (1981a) that films grown at high argon pressure have a columnar morphology and that under the described preparation conditions the coalescence of the island structure is imperfect for films grown at argon pressure higher than 20 mTorr.

In agreement with Knights (1979) and Anderson *et al.* (1979) we find that the films produced at high argon pressures have poor photoelectronic properties. As an illustration, Fig. 7 shows the photoconductivity under AM1 illumination versus the argon pressure in the discharge. The photoconductivity of the film produced at 40 mTorr of argon is lower by three orders of magnitude than that of the films produced at low argon pressures.

In order to investigate whether these structural changes are the result of bombardment or lack of bombardment of the film by neutral or charged particles we determined the dependence of the self-bias potential of the substrates on the partial pressure of argon. The substrate self-bias voltage was measured with respect to the ground in a separate experiment, which simulated the conditions of the sample deposition. Figure 8 shows the substrate self-bias potential, with respect to the ground, versus the partial pressure of argon in the discharge. The substrates are negatively biased at low argon pressures, because they are bombarded by more electrons than ions due to the higher electron mobility. At high argon pressures, on the other hand, the substrates have approximately the same potential as the plasma (positive), and charged particles do not bombard but migrate to the substrates by thermal diffusion. Ross and Messier (1981b) have measured,

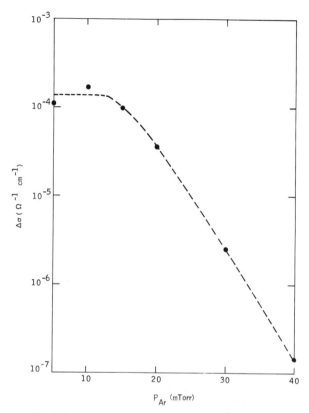

FIG. 7. Photoconductivity, measured under AM1 illumination, versus argon pressure in the discharge. $T_s = 275°C$. $P_H = 0.7$ mTorr. (After Moustakas, 1982.)

in addition, the plasma potential with a Langmuir probe and they were able to calculate the potential of the substrates with respect to the plasma. Their data have the same functional behavior observed in Fig. 8. We concluded from the data of Figs. 5–8 that the change in the sign of self-bias potential with respect to the ground occurs at exactly the same argon pressure where significant changes in the film's microstructure have been observed. Therefore, these data suggest that the coalescence of the island structure is the result of atomic rearrangement due to the bombardment of the growing film by charged particles. However, these data cannot differentiate the relative roles of electron or Ar^+ bombardment in the growth of the film. Ross and Messier (1981b) attribute the structural changes to the Ar^+ ion bombardment. However, in the next section we will show that electron bombardment may be as important in inducing structural changes. The correlations

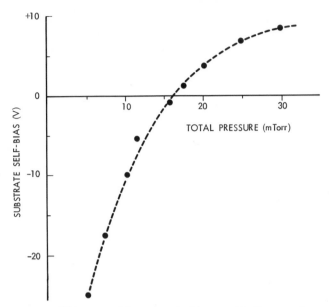

Fig. 8. Substrate self-bias potential, measured with respect to the ground versus the total pressure in the discharge. (From Moustakas, 1982.)

observed in Figs. 5–8 cannot be explained on the basis of film bombardment by neutral species, as suggested by Anderson *et al.* (1979) since such species are unaffected by substrate bias.

VI. Compositional Inhomogeneities

Compositional inhomogeneities have been observed both in glow discharge and sputtered a-Si:H films through NMR, neutron scattering, and infrared vibrational studies (Reimer *et al.,* 1980; Jeffrey *et al.,* 1981; Carlos *et al.,* 1981; Postol *et al.,* 1980; Freeman and Paul, 1978; Knights, 1979). Some authors have attempted to correlate the observed compositional inhomogeneities with the type of structural inhomogeneities that were discussed in the previous section. However, such interpretations may not be totally correct, since films with no noticeable microstructure (low argon pressure sputtered films) do show two types of different local environments for hydrogen.

Compositional inhomogeneities are expected to influence the film properties because they introduce compositional disorder. This type of disorder together with other types of disorder (e.g., thermal and structural) should in principle determine the extent of the band tails. Therefore compositional inhomogeneities are expected to affect carrier transport, optical properties, and possibly carrier recombination. In view of this, understanding the origin of such inhomogeneities and developing methods to control them is of great importance.

In sputtered mateial one type of compositional inhomogeneity is indicated by the Si–H stretching vibration, which is a doublet in the IR transmission spectrum (Brodsky et al., 1977; Freeman and Paul, 1978). Although there are multiple opinions (Paul, 1980) as to the type of SiH local environments that give rise to this doublet, it is evident that its existence indicates that these films are compositionally inhomogeneous. The inability to control the bonding of hydrogen in the sputtered material and the experience from the glow discharge that films with the 2100-cm^{-1} stretching mode generally have poorer electronic properties led to the widespread belief that the sputtered films are inferior to those produced by decomposition of silane. In this section we discuss methods to control this type of inhomogeneity and try to identify correlations between hydrogen bonding and electronic properties of the films.

One method to control the bonding of hydrogen is through bias sputtering (Moustakas, 1982). We find that the maximum effect of external bias occurs under deposition conditions that lead to substrate self-bias close to zero. We also find that positive bias favors the 2000-cm^{-1} Si–H stretching mode, while negative bias favors the 2100-cm^{-1} Si–H stretching mode. Figure 9

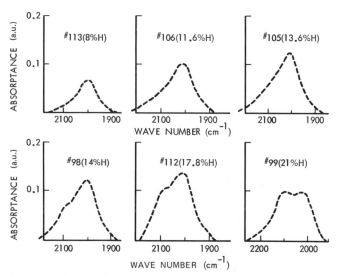

FIG. 9. Si–H stretching vibrations of films produced at positive substrate bias (+ 50 V). The hydrogen content in the film is indicated next to the sample number. (These films were produced at $T_s = 275°C$, power = 200 W, and $P_{Ar} = 15$ mTorr) (After Moustakas, 1982.)

shows the Si–H stretching vibrations of a number of films produced at substrate bias $+50$ V and at other deposition conditions as indicated in the figure caption. Indicated next to the sample number is the hydrogen content in the film. Note that most of the hydrogen, up to 14 at. %, is bonded in the 2000-cm^{-1} mode. The 2100-cm^{-1} mode acquires the same strength as the 2000-cm^{-1} mode for films having more than 20 at. % H. For contrast we show in Fig. 10 the Si–H stretching vibrations for a number of films produced under negative biasing conditions. The negative bias in these films is self-induced (-25 V), by sputtering the films at low argon pressure, $P_{Ar} = 5$ mTorr, and having all other deposition parameters the same. Note that although the hydrogen content of these films covers the same range as those of Fig. 9, the 2100-cm^{-1} mode is present even in the lowest hydrogen content films and becomes progressively stronger as the hydrogen content increases.

The structural variations associated with the substrate biasing can now be accounted for in the following way. Biasing the substrates to a positive voltage has two effects. One is electronic bombardment of the growing film; the second is minimization of Ar^+ and possibly Si^+ ion bombardment. Although electronic bombardment is thought to be undesirable in the growth of semiconductor material, our findings indicate that some electronic bombardment can have a favorable effect on the growth of structur-

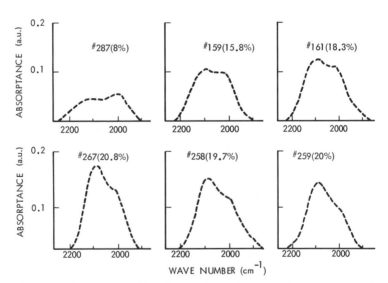

FIG. 10. Si–H stretching vibrations for a number of films produced under negative bias conditions as discussed in the text. The hydrogen content in the film is indicated next to the sample number. (After Moustakas, 1982.)

ally and compositionally homogeneous sputtered a-Si:H films. Under typical sputtering conditions, there are low-energy and high-energy electrons arriving at the substrate. Bombardment by these electrons can have profound effect on the film's chemistry and structure because it enhances adatom surface mobility through surface heating. We believe that this electron bombardment is responsible for the perfect coalescence of the island structure, which leads to films having both structural and compositional homogeneity. The second effect of the positive bias, namely, the minimization of the Ar^+ and possibly Si^+ ion bombardment, is also highly desirable. These films contain much less argon since the Ar^+ ions require a certain threshold ($\sim 15-20$ eV) in order to be implanted into the film (Comas and Carosella, 1968). Resputtering is also minimized under these conditions of film growth. Since resputtering by Ar^+ ions is likely to be anisotropic (i.e., with higher sputtering yield at the low-density intercolumnar regions), there is a high possibility of microvoid formation during the film regrowth in these regions. Both argon incorporation and void formation lead to films with structural and compositional inhomogeneity. In fact, there is some evidence that argon inhabits voids (Paul et al., 1973). We believe that such structural inhomogeneities result in the formation of a stretching vibration doublet.

We now examine whether there is a correlation between the type of Si–H bonding and the electronic properties of the films. Such correlations were sought by studying the density of states in the middle of the gap, the photoconductivity, and the optical absorption constant.

The infrared data in Fig. 10 were measured on the films whose densities of states in the middle of the gap are given in Fig. 1. Note that the sample with the lowest density of states (7×10^{14} cm^{-3} eV^{-1}) has the most of its hydrogen bonded in the 2100-cm^{-1} mode. From these data we concluded that the states in the middle of the gap and properties that are related to these states (recombination) do not depend on the type of Si–H bonding but rather on the total amount of bonded hydrogen.

Next we compare the optical properties of films that have the Si–H stretching vibration primarily at 200 cm^{-1} with others in which the 2100-cm^{-1} mode is the dominant one. For this study we deposited two series of films; in one series the substrates were electrically biased at +50 V and in the other were electrically grounded. The hydrogen partial pressure in each series was adjusted so as to produce films with optical gaps in the range of 1.6–1.9 eV. All other deposition parameters were fixed, namely, the deposition temperature was 275°C, the argon pressure 15 mTorr, and the power in the discharge 1.5 W cm^{-2}. The hydrogen content in these films was determined by the ^{15}N nuclear reaction (Moustakas et al., 1981a) and the Si–H bonding configuration was determined by studying the IR vibrational spec-

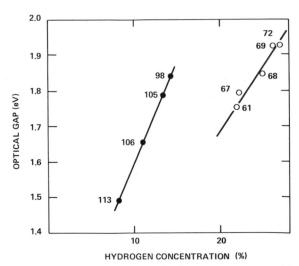

FIG. 11. Optical gap versus hydrogen content for films produced on positively biased and electrically grounded substrates. [The optical gap E_g was defined by the relation $(\alpha h\nu)^{1/2} = B(h\nu - E_g)$.] ●, substrates positively biased; ○, substrates electrically grounded. (After Moustakas, 1983.)

tra. The stretching vibration spectrum for the biased samples centers at 2000 cm^{-1}, while that for the grounded samples is a doublet with the 2100-cm^{-1} mode becoming progressively stronger as the hydrogen content increases. The optical gap of these films versus the hydrogen content in the films is shown in Fig. 11. These data clearly indicate that structurally and compositionally inhomogeneous films require more hydrogen to attain the same optical gap. The fact that the more homogeneous films have larger gaps for the same amount of hydrogen is consistent with the model proposed by Cody et al. (1981a,b) that the optical gap is related to disorder. Our data suggest that in addition to structural and thermal disorder, which was proposed by Cody, compositional disorder is also important in determining the optical properties of a-Si:H alloys.

The studies we presented in this section suggest that compositional inhomogeneities in films with no noticeable microstructure do not affect the states in the middle of the gap, but they do affect the band tails through the introduction of additional disorder.

VII. Device Potential

The recent progress in fabricating high-efficiency solar cells by sputtering (Moustakas, 1983a,b) suggests that sputtered a-Si:H films are not inferior in their electronic properties to those produced by the glow discharge process.

Electron bombardment, which is generally unacceptable in crystalline surface sensitive devices, can be controlled and used to facilitate the coalescence of the island structure during the growth of the film.

These findings, together with the fact that sputtering is a widely practiced fabrication technique with an extensive compendium of art, developed skills, and available hardware, suggest that this method will play a significant role in this emerging technology of amorphous materials. It should be pointed out that continuous processing plants using sputtering technology have proved successful for other materials. There is no intrinsic limit to size; scaling up increases deposition rates and improves film uniformity. The method is suitable also for depositing other amorphous semiconductors [such as Ge (Lewis *et al.*, 1974) and GaAs (Paul *et al.*, 1977)] or their alloys and for forming chemically graded junctions and tandem solar cells (Moustakas, 1983b).

VIII. Future Research

In this review we presented data and models of interpretations of the different possible mechanisms of film growth. We also show that low-energy ion and electron bombardment of the growing film during deposition affects the nucleation kinetics and improves structural and compositional inhomogeneities. The models of film growth, particularly the possibility that these films grow by a combination of physical sputtering and glow discharge decomposition of in situ generated silane, need to be tested and refined. The electron and ion bombardment effects need to be studied in more detail and optimized for best results.

Another area that needs more research is the phenomenon of substitutional doping. Sputtering is a suitable technique for this type of study because it can easily explore interactions between dopant atoms, natural defects, and hydrogen atoms.

The concept of hydrogen insertion into a reconstructed dangling bond (a bridging H shared by two neighboring Si) should be tested by searching for its characteristic spin resonance that should differ from that of a single dangling bond.

Finally, other types of sputtering such as cylindrical magnetron (Thornton and Hoffman, 1981), planar magnetron (Stein *et al.*, 1981), and ion beam sputtering (Kobayashi *et al.*, 1981) need to be explored in more detail.

ACKNOWLEDGMENTS

The author gratefully acknowledges the partial support of the U.S. Department of Energy (Solar Energy Institute Contract No. ZB-3-02166-01) during the course of preparation of this chapter.

REFERENCES

Anderson, D. A., Moustakas, T. D., and Paul, W. (1977). *In* "Amorphous and Liquid Semiconductors" (W. E. Spear, ed.), pp. 334–338. University of Edinburgh, Scotland.

Anderson, D. A., Moddel, G., Paesler, M. A., and Paul, W. (1979). *J. Vac. Sci. Technol.* **16,** 906.

Anderson, D. A., Moddel, G., and Paul, W. (1980). *J. Electron. Mater.* **9,** 141–152.

Biegelsen, D. K., Street, R. A., Tsai, C. C., and Knights, J. C. (1979). *Phys. Rev. B* **20,** 4839.

Brodsky, M. H., Title, R. S., Weiser, K., and Pettit, G. D. (1970). *Phys. Rev. B: Solid State* [3] **1,** 2632–2641.

Brodsky, M. H., Cardona, M., and Cuomo, J. J. (1977). *Phys. Rev. B: Solid State* [3] **16,** 3556.

Bruyere, J. C., Deneuville, A., Mini, A., Fontenille, J., and Danielou, R. (1980). *J. Appl. Phys.* **51,** 2199.

Carlos, W. E., Taylor, P. C., Oguz, S., and Paul, W. (1981). *In* "Tetrahedrally Bonded Amorphous Semiconductors" (R. A. Street, D. K. Biegelsen, and J. C. Knights, eds.), pp. 67–72. Am. Inst. Phys., New York.

Catalano, A., D'Aiello, R. V., Dresner, J., Faughnan, B., Firester, A., Kane, J., Schade, H., Smith, Z. E., Swartz, G., and Triano, A. (1982). *Conf. Rec. IEEE Photovoltaic Spec. Conf.* **16,** 1421–1422.

Coburn, J. W., and Kay, E. (1972). *J. Appl. Phys.* **43,** 4965.

Cody, G. D., Tiedje, T., Abeles, B., Moustakas, T. D., Brooks, B., and Goldstein, Y. (1981a). *J. Phys. (Orsay, Fr.)* **42,** *Suppl.* **10,** C4-301–C4-304.

Cody, G. D., Tiedje, T., Abeles, B., Brooks, B., and Goldstein, Y. (1981b). *Phys. Rev. Lett.* **47,** 1480.

Comas, J., and Carosella, C. A. (1968). *J. Electrochem. Soc.* **115,** 974.

Connell, G. A. N., and Pawlik, J. R. (1976). *Phys. Rev. B* **13,** 787.

Freeman, E. C., and Paul, W. (1978). *Phys. Rev. B* **18,** 4288.

Heller, J. (1973). *Thin Solid Films* **17,** 163.

Jeffrey, F. R., Shanks, H. R., and Danielson, G. C. (1979). *J. Appl. Phys.* **50,** 7034–7038.

Jeffrey, F. R., Shanks, H. R., and Danielson, G. C. (1980). *J. Non-Cryst. Solids* **35–36,** 261–266.

Jeffrey, F. R., Lowry, M. E., Garcia, M. L. S., Barnes, R. G., and Torgeson, D. E. (1981). *In* "Tetrahedrally Bonded Amorphous Semiconductors" (R. A. Street, D. K. Biegelsen, and J. C. Knights, eds.), pp. 83–88. Am. Inst. Phys., New York.

Knights, J. C. (1979). *Jpn. J. Appl. Phys.* **18,** 101.

Knights, J. C., and Lujan, R. A. (1979). *Appl. Phys. Lett.* **35,** 244.

Kobayashi, M., Saraie, J., and Matsunami, H. (1981). *Appl. Phys. Lett.* **38,** 696–697.

Lewis, A. J. (1976). *Phys. Rev. B* **14,** 658–668.

Lewis, A. J., Connell, G. A. N., Paul, W., and Pawlik, J. R. (1974). *In* "Tetrahedrally Bonded Amorphous Semiconductors" (M. H. Brodsky and S. Kirkpatrick, eds.), pp. 27–31. Am. Inst. Phys., New York.

Maissel, L. I. (1970). *In* "Handbook of Thin Film Technology" (L. I. Maissel and R. Glang, eds.), Chapter 4. McGraw-Hill, New York.

Martin, P. M., and Pawlewicz, W. T. (1981). *J. Non-Cryst. Solids* **45,** 15–27.

Maruska, H. P., Moustakas, T. D., and Hicks, M. C. (1983). *Sol. Cells* **9,** 37–51.

Matsuda, A., Nakagawa, K., Tanaka, K., Matsumura, M., Yamasaki, S., Okushi, H., and Iizima, S. (1980). *J. Non-Cryst. Solids* **35–36,** 183–188.

Morel, D. L., and Moustakas, T. D. (1981). *Appl. Phys. Lett.* **39,** 612–614.

Moss, S. C., and Graczyk, J. F. (1969). *Phys. Rev. Lett.* **23,** 1167.

Moustakas, T. D. (1979). *J. Electron. Mater.* **8,** 391–435.

Moustakas, T. D. (1982). *Sol. Energy Mater.* **8,** 187–104.

Moustakas, T. D. (1983a). *In* "Photovoltaics for Solar Energy Applications II" (D. Adler, ed.), *Proc. SPIE* **407**, 56–64.

Moustakas, T. D. (1983b). *Proc. EC Photovoltaic Conf., 5th*

Moustakas, T. D., and Connell, G. A. N. (1976). *J. Appl. Phys.* **47**, 1322–1326.

Moustakas, T. D., and Friedman, R. (1982). *Appl. Phys. Lett.* **40**, 515–517.

Moustakas, T. D., and Maruska, H. P. (1983). *Appl. Phys. Lett.* **43**, 1037.

Moustakas, T. D., and Paul, W. (1977). *Phys. Rev. B: Solid State* [3] **15**, 1564–1576.

Moustakas, T. D., Anderson, D. A., and Paul, W. (1977). *Solid State Commun.* **23**, 155–158.

Moustakas, T. D., Wronski, C. R., and Morel, D. L. (1980). *J. Non-Cryst. Solids* **35–36**, 719–724.

Moustakas, T. D., Tiedje, T., and Lanford, W. A. (1981a). *In* "Tetrahedrally Bonded Amorphous Semiconductors" (R. A. Street, D. K. Biegelsen, and J. C. Knights, eds.), pp. 20–24. Am. Inst. Phys., New York.

Moustakas, T. D., Wronski, C. R., and Tiedje, T. (1981b). *Appl. Phys. Lett.* **39**, 721–723.

Moustakas, T. D., Friedman, R., and Weinberger, B. R. (1982). *Appl. Phys. Lett.* **40**, 587–588.

Moustakas, T. D., Maruska, H. P., Friedman, R., and Hicks, M. (1983). *Appl. Phys. Lett.* **43**, 368–370.

Movchan, B. A., and Demshishin, A. V. (1969). *Fiz. Met. Metalloved.* **28**, 653.

Oguz, S. (1981). Thesis, Harvard University, Cambridge, Massachusetts (unpublished).

Oguz, S., and Paesler, M. A. (1980). *Phys. Rev. B* **22**, 6213.

Oguz, S., Collins, R. W., Paesler, M. A., and Paul, W. (1980). *J. Non-Cryst. Solids* **35–36**, 231.

Oguz, S., Paul, D. K., Blake, J., Collins, R. W., Lachter, A., Yacobi, B. G., and Paul, W. (1981). *J. Phys. (Orsay, Fr.)* **42**, *Suppl.* **10**, C4-679–C4-82.

Paesler, M. A., Okumura, T., and Paul, W. (1980). *J. Vac. Sci. Technol.* **17**, 1332.

Paul, W. (1980). *Solid State Commun.* **34**, 283–285.

Paul, W., and Anderson, D. A. (1981). *Sol. Energy Mater.* **5**, 229–316.

Paul, W., Connell, G. A. N., and Temkin, R. J. (1973). *Adv. Phys.* **22**, 529–580.

Paul, W., Lewis, A. J., Connell, G. A. N., and Moustakas, T. D. (1976). *Solid State Commun.* **20**, 969–972.

Paul, W., Moustakas, T. D., Anderson, D. A., and Freeman, E. (1977). *In* "Amorphous and Liquid Semiconductors" (W. E. Spear, ed.), pp. 467–471. University of Edinburgh, Scotland.

Pawlewicz, W. T. (1978). *J. Appl. Phys.* **49**, 5595–5601.

Postol, T. A., Falco, C. M., Campwirth, R. T., Schuller, I. K., and Yelon, W. B. (1980). *Phys. Rev. Lett.* **45**, 648.

Reimer, J. A., Vaughan, R. W., and Knights, J. C. (1980). *Phys. Rev. Lett.* **44**, 193.

Ross, R. C., and Messier, R. (1981a). *J. Appl. Phys.* **52**, 5329–39.

Ross, R. C., and Messier, R. (1981b). *In* "Tetrahedrally Bonded Amorphous Semiconductors" (R. A. Street, D. K. Biegelsen, and J. C. Knights, eds.), pp. 53–57. Am. Inst. Phys., New York.

Schwartz, N. (1964). *Trans. Natl. Vac. Symp.* **10**, 325.

Shanks, H., Fang, C. J., Ley, L., Cardona, M., Demond, F. J., and Kalbitzer, S. (1980). *Phys. Status Solidi B* **100**, 43.

Shevchik, N. J., and Paul, W. (1972). *J. Non-Cryst. Solids* **8–10**, 381.

Spear, W. E. (1974). *In* "Amorphous and Liquid Semiconductors" (J. Stuke and W. Brenig, eds.), pp. 1–16. Taylor & Francis, London.

Spear, W. E., and LeComber, P. G. (1975). *Solid State Commun.* **17**, 1193–1196.

Stein, H. J., Peercy, P. S., and Peckerar, M. (1981). *J. Electron. Mater.* **10**, 797.

Temkin, R. J., Paul, W., and Connell, G. A. N. (1973). *Adv. Phys.* **22**, 581.

Thompson, M. J., Allison, J., Al-Kaisi, M. M., and Thomas, I. P. (1978). *Rev. Phys. Appl.* **13**, 625–628.

Thornton, J. A. (1974). *J. Vac. Sci. Technol.* **11,** 666–670.

Thornton, J. A., and Hoffman, D. W. (1981). *J. Vac. Sci. Technol.* **18,** 203–207.

Tiedje, T., Moustakas, T. D., and Cebulka, J. M. (1981). *Phys. Rev. B* **23,** 5634–5637.

Turner, D. P., Thomas, I. P., Allison, J., Thompson, M. J., Rhodes, A. J., Austin, I. G., and Searle, T. M. (1981). *In* "Tetrahedrally Bonded Amorphous Semiconductors" (R. A. Street, D. K. Biegelsen, and J. C. Knights, eds.), pp. 47–51. Am. Inst. Phys., New York.

Vepřek, S. (1980). *Chimia* **34,** 489.

Victorovitch, P., Jousse, D., Chenevas-Paule, A., and Lieuz-Rocherz, L. (1979). *Rev. Phys. Appl.* **14,** 204.

von Roedern, B., Ley, L., Cardona, M., and Smith, F. W. (1979). *Philos. Mag [Part] B* **40,** 433–450.

Vossen, J. L., and Cuomo, J. J. (1978). *In* "Thin Film Processes" (J. L. Vossen and W. Kern, eds.), Chapter 2. Academic Press, New York.

Vossen, J. L., and Kern, W., eds. (1978). "Thin Film Processes." Academic Press, New York.

Wolski, S. P., and Zdanuk, E. J. (1961). *J. Appl. Phys.* **32,** 782.

CHAPTER 5

Ionized-Cluster Beam Deposition

Isao Yamada

ION BEAM ENGINEERING EXPERIMENTAL LABORATORY
KYOTO UNIVERSITY
SAKYO, KYOTO, JAPAN

I. Introduction

In the ionized-cluster beam (ICB) deposition, ions of macroaggregated atoms (clusters) are utilized instead of ions of individual atoms. The idea and the first data related to the generation of vaporized clusters and the formation of films by ionized-cluster beam were reported by Takagi *et al.* in 1972 (See Takagi *et al.*, 1972). The term *cluster,* in general, means an aggregate of atoms or molecules, whose size ranges from dimer and trimer up to 10^5 or more atoms. Our interest is mainly concentrated in the "vaporized-material clusters" that consist of several hundreds to several thousands of atoms. In such clusters a considerable portion of constituent atoms are located on its surface. Consequently, the physical properties of such clusters are quite different form those of liquid-state droplets or of solids (Stein, 1979).

According to the classical theory of droplets, a vaporized-material cluster consisting of several hundreds to several thousands of atoms of solid-state material would be difficult to form, assuming the surface tension of bulk liquid to apply. However, a recent theory predicts that the surface tension tends to decrease as the size of the droplet decreases, and the surface tension of the bulk liquid may not be applied to clusters containing less than several thousands of atoms because their surface layer cannot be clearly defined

83

(Hirschfelder *et al.*, 1964; Kristensen *et al.*, 1974). Furthermore, a study by Monte Carlo simulation suggests that the clusters have no definable surface (Lee *et al.*, 1973).

In the ICB technique, clusters that consist of 500–2000 loosely coupled atoms can be formed by ejecting the vapor of many kinds of solid material (metals, semiconductors, and insulators) through the nozzle of a heated crucible into a high-vacuum region. The clusters ejected from the crucible are ionized by electron bombardment in an ionization chamber and are subsequently accelerated toward the substrate by a high negative potential. Accelerated ionized clusters and neutral clusters that were not accelerated are deposited on the substrate surface. One of the most important features of the cluster beam deposition is that the ionized clusters can transport what is equivalent to a high current at very low energy (Takagi *et al.*, 1975); it also uses properties inherent to the cluster (Yamada *et al.*, 1982a), i.e, their kinetic energy and their ionic charge (Takagi, 1982a). By controlling the kinetic energy and the proportion of clusters that are ionized, film formation kinetics can be controlled. In the ICB deposition of amorphous silicon (a-Si: H) film, the deposition is made using a partially ionized silicon cluster beam in hydrogen gas at pressures on the order of $10^{-5}-10^{-4}$ Torr.

In this chapter the formation and the properties of the cluster beam and the film formation mechanism by ICB are discussed. Then apparatus and conditions to form a-Si: H films and resultant film characteristics are described.

II. Formation and Properties of the Clusters

A cluster source that can be used in the ICB deposition is a cylindrical nozzle source. The dimensions and other factors in the design of the cluster-forming source are different from those of a molecular beam source. Figure 1 and Table I show the conditions of source construction and operation where D is the nozzle diameter, λ the mean free path of the atoms in the crucible, P_0 is the inner pressure of the crucible, P the vapor pressure of the beam outside the crucible, and L the thickness of the nozzle.

With regard to cluster formation (Yamada *et al.*, 1982b), the atoms of

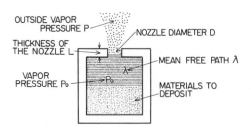

FIG. 1. Dimensions and parameters of the cluster source.

TABLE I

Conditions of Source Construction
and Operation[a]

	Examples
$D \gg \lambda$	$D = 0.1 - 2$ mm, $\lambda = 0.005 - 0.5$ mm
	(multiple nozzle is available)
$P_0/P \gtrsim 10^4 - 10^5$	$P = 10^{-7} - 10^{-4}$ Torr,
	$P_0 = 10^{-2}$ - several Torr
$L/D = 0.5 - 2.0$	For cylindrical nozzle, experimentally

[a] Supersaturated vapor in adiabatic expansion

depositing material collide and transfer their energies to each other during the ejection through the nozzle of the crucible into a region of higher vacuum. Then a supersaturated state is produced by adiabatic expansion. The atoms, which lose their energy by collision, start to aggregate to form nuclei. Nuclei smaller than a critical size are not stable and break up into smaller pieces. However, the nuclei formed in the supersaturated state that are larger than the critical size grow to form clusters. The growth of the clusters occurs at a maximum rate near the nozzle and then slows down and finally ceases in the region where the pressure decreases and the collision frequency becomes lower. The formation of a cluster is initiated by homogeneous condensation of the supersaturated vapor in an adiabatic expansion and the clusters grow by collisions with the surrounding vaporized atoms in the nozzle region, where the nozzle diameter D has to be larger than the mean free path λ of the atoms in the crucible. To make the adiabatic expansion more effective, the nozzle diameter is sufficiently small to satisfy the condition that the ratio of the inner pressure P_0 of the crucible to the vapor pressure P outside the crucible is greater than 10^4. It is also desirable for D to be comparable with the nozzle thickness L, that is, $L/D = 0.5 - 2.0$ according to experimental results. In a high vacuum region the size of most of the cluster remains constant because there is little collision between clusters and gas atoms.

Figure 2 shows an example of energy analysis of an ionized-cluster beam as a function of the vapor pressure P_0 of the depositing material in the crucible (Yamada and Takagi, 1981). Two kinds of peaks are observed in the spectra. Weak peaks of close to zero intensity correspond to monoatomic particles and strong peaks near $80 - 120$ eV correspond to the clusters. The intense peak from the clusters appear in that pressure range where the formation conditions as shown in Table I are satisfied. By combining the results with ejection velocity measurements made using a rotating-disk method, the cluster size has been calculated to be $500 - 2000$ atoms per cluster.

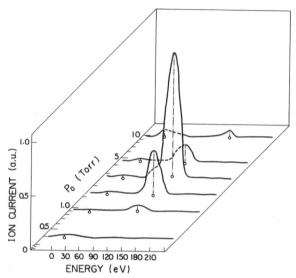

FIG. 2. Energy distribution of the cluster beam as a function of the vapor pressure in the crucible. Nozzle diameter is 1 mm; nozzle length is 1 mm. Ag was used as a test material.

Typical operating conditions recommended in practical use are $P_0 = 10^{-2}$ to several Torr, since the vapor pressure P in the chamber is about $10^{-7} - 10^{-4}$ Torr, and $D = 0.5 - 2.0$ mm (note that single-nozzle and multiple-nozzle sources can be used). In our experiment the nozzle diameter D was 1 mm and $L/D = 1$. At a pressure P_0 higher than 5 Torr the intensity of the cluster beam begins to decrease. This is considered to be because the difference in pressures inside and outside the crucible decreases owing to the high rate of ejection from the crucible. The maximum and minimum values of P_0 can be extended by changing the shape or the dimensions of the nozzle. Figure 3 shows the influence of the nozzle diameter on the energy and intensity of the clusters. The mean free path of the atoms in the vapor under the experimental conditions of 1.25 Torr is of the order of 0.01 mm, and therefore the diameters of all the nozzles tested are sufficiently large for the occurrence of the many collisions necessary for cluster growth during the expansion. In the case of the 2-mm nozzle, clusters were formed most effectively. However, no peak at higher energy is observed for a crucible with a diameter larger than 5 mm, including an open crucible, because the pressure difference between the inside and the outside of the crucible is insufficient. Under these conditions that are similar to the case of the conventional evaporator, the vapor consists of atoms, molecules, and/or small clusters composed of several atoms instead of larger clusters.

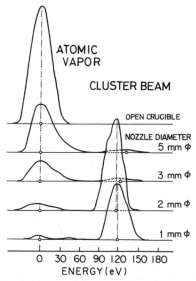

FIG. 3. Energy distribution of the cluster ions for different nozzle diameters. $P = 1.25$ Torr. Crucible inner diameter is 8 mm ϕ.

A relation between the intensity of the cluster and the vapor pressure in the crucible at various nozzle diameters is shown in Fig. 4. Taking the values of $P_{0,max}$ and D for different nozzle dimensions, where $P_{0,max}$ is the optimum pressure that gives the maximum intensity, it was shown that the product $P_{0,max} \cdot D$ was approximately constant. Since λ is inversely proportional to

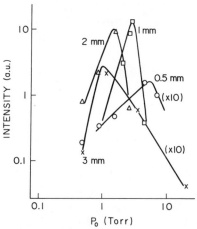

FIG. 4. Relation between beam intensity and vapor pressure in the crucible with different nozzle diameters labeling each curve. Nozzle length is 1 mm.

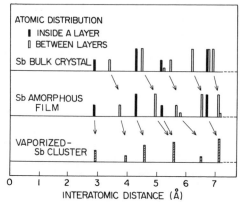

Fig. 5. Comparison of the interatomic distances in various kinds of states. Sb was used as a source material.

P_0, the preceding relation means that the number of collision D/λ during the expansion is not changed by the nozzle dimensions. These results also suggest that the clusters can be formed during homogeneous condensation by adiabatic expansion (Hagena, 1974).

The microscopic structure of the cluster can be examined by an electron diffraction analysis and compared with that of bulk and amorphous thin films (Yamada *et al.*, 1982a). The electron diffraction data were used to calculate the distance between atoms in the cluster. Figure 5 shows a comparison of the interatomic distances in a cluster with those of bulk crystal and amorphous films. The interatomic distance in the amorphous film is longer than that in the crystal. A further spread in the interatomic distance can be seen in our clusters. This characteristic can be used to explain unique film formation kinetics in the ICB deposition process.

III. ICB Deposition Equipment

A typical schematic diagram of the ionized-cluster beam deposition system for a-Si:H film formation is shown in Fig. 6. The ICB deposition system consists of three sections: cluster beam generation, ionization, and acceleration. In the cluster beam source the crucible is heated either by electron bombardment or by radiation or direct resistance heating. In Fig. 6 the electron bombardment type is shown. The electron bombardment method is most suitable for the deposition of high-temperature materials. For large-scale industrial use a hybrid system of electron bombardment and direct resistive heating is convenient (Takagi *et al.*, 1981). The clusters are ionized to be singly charged by electron bombardment in the ionization section located above the crucible. The ratio of ionized clusters to total

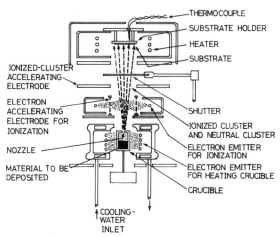

THERMOCOUPLE

SUBSTRATE HOLDER

HEATER

SUBSTRATE

IONIZED-CLUSTER ACCELERATING ELECTRODE

ELECTRON ACCELERATING ELECTRODE FOR IONIZATION

SHUTTER

IONIZED CLUSTER AND NEUTRAL CLUSTER

NOZZLE

ELECTRON EMITTER FOR IONIZATION

MATERIAL TO BE DEPOSITED

ELECTRON EMITTER FOR HEATING CRUCIBLE

CRUCIBLE

COOLING - WATER INLET

FIG. 6. ICB deposition system for the deposition of a-Si films.

clusters can be adjusted by changing the ionizing electron current I_e. For example, the degree of ionization is 5–7% at $I_e = 100$ mA, 7–15% at $I_e = 150$ mA, and 30–35% at $I_e = 300$ mA (Takagi et al., 1978a). To obtain a uniform spatial distribution of the ion beam on the substrate surface a square electron acceleration cage is used (Takagi et al., 1978b). The electron acceleration cage is constructed of parallel wires through which electrons emitted from a filament can be accelerated toward the center of the cage. The ionized clusters formed in this section are accelerated to the substrate surface by an electric field in the acceleration section.

For hydrogenated amorphous silicon film formation, the deposition is carried out in hydrogen gas in a range of $10^{-5} - 10^{-4}$ Torr. The hydrogen gas is introduced into the chamber through a variable leak valve. For the deposition of doped films such as p- or n-type films, one uses diborane or phosphine in a hydrogen carrier gas with a dilution of a few thousand ppm. The system construction is shown in Fig. 7. The pressure of three different gases is controlled independently by monitoring the gas pressure in the chamber. In the ICB deposition of doped a-Si:H, different conductivity types of film can be deposited in the same chamber by changing the doping gas without cross-contamination by unwanted residual gases in the chamber or by gas desorbed from the chamber walls. The ICB deposition method is essentially an ion deposition technique where sustaining a plasma is not necessary and where the reactive gas pressure is always maintained below the pressure at which a plasma would be produced. Therefore, undesirable gas inclusion into the film, the so-called gas precipitation, can be eliminated. If a plasma were produced in the chamber, adsorbed gases on the inside wall

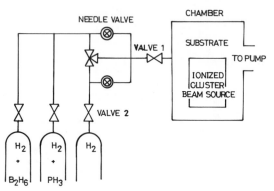

FIG. 7. Gas system feeding the deposition chamber.

would be desorbed and would contaminate the substrate. Moreover, clusters would be destroyed by plasma heating and the advantages of the ICB technique would be lost.

In Fig. 8 a view of a commercial production system is shown (Takagi, 1982b). The equipment is a one-crucible system. Three extra crucibles can be mounted in the chamber and can be interchanged succesively by remote control. Deposition conditions such as crucible temperature, ionization current and voltage, acceleration voltage, hydrogen gas pressure, deposition rate, substrate temperature, film thickness, and so on, can be set by an

FIG. 8. ICB commercial deposition system suitable for amorphous silicon deposition.

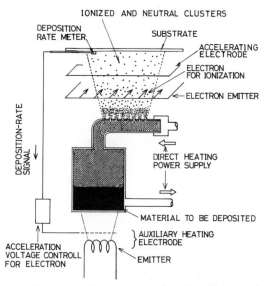

FIG. 9. ICB source with single crucible and multiple nozzles.

automatic control system. Figure 9 shows a single-crucible system with multiple nozzles to form a ribbon beam for producing a wide sheet beam having a high uniformity to coat large substrates on an industrial scale (Takagi *et al.*, 1981).

IV. Film Formation Mechanism

Generally, the analysis of deposition features in ion-assisted film formation such as plasma deposition, ion plating, and ion beam sputter deposition is extremely complex. But in the ICB deposition the influences of the deposition parameters on film growth can be easily estimated by measuring deposition conditions because the deposition is carried out in a relatively high-vacuum region. In the actual deposition the film quality is considered to be influenced by the following conditions: (1) the impinging rate of the depositing particles, (2) the substrate temperature, (3) the ambient gas or reactive gas pressure, (4) the substrate surface condition, (5) the kinetic energy of depositing particles, (6) the content of ions in the total depositing particles. Effects of the kinetic energy and of the ionic charge of the ionized clusters are summarized in Table II.

Each atom of the cluster has an average energy $E = e \cdot V_a/N$, where e is the electronic charge, V_a is the acceleration voltage, and N is the number of atoms per cluster. By controlling V_a, it is possible to provide each atom with energies high enough for surface diffusion ($E \sim 1$ eV), Surface cleaning by

TABLE II

EFFECTS OF IONIZED CLUSTERS ON FILM FORMATION

A. Effects due to the extremely large mass-to-charge ratio
  1. Equivalent high current transport
  2. Control of the average particle energy in the ionized clusters over the range from thermal ejection to above 100 eV within which optimum conditions for film growth are generally achieved
B. Effects due to the kinetic energy of the ionized clusters
  1. Sputtering of absorbed impurities or contaminations from the substrate surface and from the deposited surface
  2. Heating at equivalently high temperature
  3. Implantation of deposits into extremely shallow part of the substrate surface
  4. Adatom migration on the substrate surface
  5. Creation of activated centers for nucleation
C. Effects due to the charge of the ionized clusters
  1. Change of critical parameters in the condensation process during film formation such as nucleation, coalescence, and growth
  2. Chemical reaction

sputtering surface contaminants ($E = 0.1 - 10$ eV) can be achieved by adjusting the acceleration voltage at relatively low substrate temperatures. Furthermore, sufficient energy for inducing a suitable number of defects and atomic displacements in the layer ($E > 10$ eV) can be provided, if necessary.

These fundamental effects during film formation by ICB deposition can be observed in early stages of the film formation. Nucleation density and adatom migration and sputtering effects were observed by changing the deposition conditions. The density of nucleation centers increases monotonically at first as the acceleration voltage is increased and then it saturates (Takagi, 1982c). The adatom migration effect was investigated by electron microscopy of the films (Takagi *et al.,* 1976). A considerably large adatom migration due to the imparted momentum until they are trapped by the nuclei was observed as a function of the acceleration voltage. The loosely coupled atoms within the cluster can easily break up into free atoms upon impacting the substrate surface where upon they can diffuse over the surface. When the ionized clusters are accelerated to the substrate surface, their energy is converted to surface diffusion energy. The migration distance increases with increasing acceleration voltage. The migration effect of adatoms due to this enhanced surface diffusion energy contributes to the formation of good-quality films or to changing the morphology of the deposited films.

These effects are clearly seen by comparing the depositions of epitaxial silicon films in high vacuum in the range of $10^{-7} - 10^{-6}$ Torr and in ultra-high vacuum better than 10^{-10} Torr (Takagi, 1982c). In silicon epitaxial films made in a high-vacuum chamber that was evacuated using an oil diffusion pump to operate at a base pressure of 10^{-7}–10^{-6} Torr, only a standard chemical cleaning procedure was used to clean the substrate before the deposition. The diffraction pattern of the film using a 100-kV electron beam showed that the film deposited at lower acceleration voltage was amorphous or polycrystalline. But an epitaxial film could be obtained at an acceleration voltage higher than 6 kV at a substrate temperature in the 400–630°C range (Yamada et al., 1980a). In the case of deposition in an ultra-high-vacuum chamber, an atomically clean and well-ordered silicon surface was prepared as a substrate. Then the ICB deposition of silicon on the clean surface was made at a substrate temperature of 500°C. The film formed in this method was examined by in situ electron diffraction at 5 kV. The comparison of the diffraction patterns is shown in Fig. 10. In the high-vacuum case the deposited film was influenced by the native oxide on the substrate surface. In this case it was necessary to apply a high voltage to accomplish sufficient surface cleaning. On the other hand, in the deposition on an atomically clean surface, a single crystalline pattern was observed under 200-V acceleration at a 500°C substrate temperature. By increasing the acceleration voltage, an improvement of the crystalline quality could be observed. The improved crystallization seems to be due to a change in the deposition kinetics by accelerated cluster ions.

The effects of the acceleration voltage on the film formation kinetics by ICB can be seen by measuring the mass deposited at different acceleration voltages. The mass condensed at a time t on the substrate as a function of substrate temperature with the impingement rate $\dot{M}$ of the depositing particles is given by (Neugebauer, 1970)

$$M = \dot{M}t[1 - (N_0/I^*t)\exp{(-U/kT)}],$$

where $U = \phi_{ad} - \phi_d$, I^* is the rate of formation of critical nuclei, N_0 is the density of adsorption sites on the substrate surface, ϕ_{ad} is the activation energy for desorption and ϕ_d is the activation energy for surface diffusion. Figure 11 shows the preceding characteristics as a function of the mass deposited versus reciprocal substrate temperature during silicon deposition by ICB. The figure also shows the transition temperature to crystalline properties for different acceleration voltages. Signs a, p, and s in the figure indicate amorphous, poly-, and single-crystalline states, respectively.

Since ϕ_{ad} is generally larger than ϕ_d, the curve of mass versus reciprocal temperature will have a positive slope. In the deposition by un-ionized clusters, which is similar to that of conventional deposition except that the cluster state of the particles is used, the deposited mass increases with decreasing substrate temperature. The mass deposited at a given substrate temperature decreases with increasing acceleration voltage because of sput-

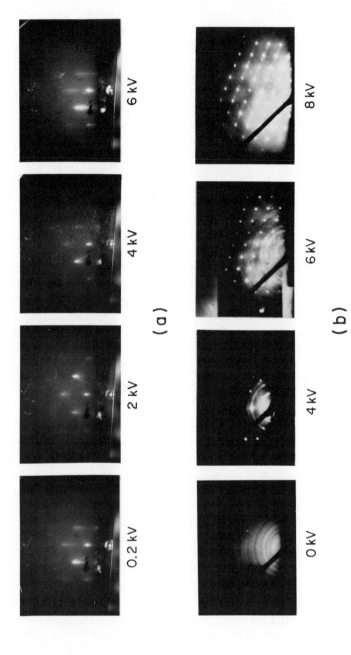

FIG. 10. Electron diffraction patterns of epitaxial silicon films by ICB at different acceleration voltages. (a) Under ultrahigh-vacuum (5×10^{-9} Torr) condition. (b) Under high-vacuum (1×10^{-6} Torr) condition.

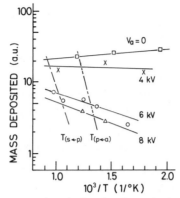

FIG. 11. Relation between mass-deposited and reciprocal substrate temperature, and the change of transition temperature for different acceleration voltages. Impinging rate $\dot{M} = 300$ Å min^{-1}.

tering. Moreover, the slope of the line showing the deposited mass changes with increasing acceleration voltage. The change of the slope from positive to negative with increasing acceleration voltage is considered to be due to a change of values such as N_0, I^*, and U. The density of adsorption sites N_0 can be controlled by creating preferential desorption sites on the substrate surface by bombardment with the ICB. The rate of formation of critical nuclei I^* can be changed by changing the beam energy. The energy associated with differential energy for deposition, U, is related to the activation energies for surface diffusion and for desorption. Since the degree of adatom migration can be changed by changing the acceleration voltage, it can be said that it is the same as if the activation energy for surface diffusion were controlled by changing the acceleration voltage, which results in a change of U. In the conventional ion beam deposition method that uses atomic

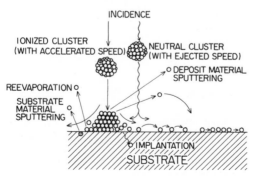

FIG. 12. A schematic diagram of film formation by the ICB.

neutral or ionized particles, the energy U does not change (Babaev *et al.*, 1976).

In the ICB deposition, unlike the conventional vacuum deposition and sputter deposition, the film structure (such as single, poly, and amorphous) can be chosen by changing the deposition conditions under relatively low substrate temperature. As discussed earlier, the process of film formation by ICB can be conceptually shown in Fig. 12.

V. Deposition Conditions

Deposition conditions of a-Si : H film are listed in Table III. The crucible is heated by electron bombardment to 2000°C. Vapor pressure in the crucible at this temperature is about 1 Torr. Typical values of applied voltage and current to heat the crucible were 1.7 kV and 0.7 A, respectively. The crucible is well shielded thermally to prevent heating the substrate surface and other components in the system as well as the walls of the chamber. An ionization voltage of 150 V was applied to the ionizing-electron beam because the maximum ionization efficiency by electron impact is in the range of 100–200 eV. Hydrogen gas was introduced through a variable leak valve to the chamber. The base pressure before the introduction of hydrogen gas was 5×10^{-7} Torr.

Figure 13 shows the dependence of the deposition rate on the hydrogen gas pressure. The deposition rate falls with increasing hydrogen gas pressure. Taking into account the mean free path of hydrogen molecules, a silicon cluster is not expected to collide with a hydrogen molecule in its path to the substrate. Therefore, the reduced deposition rate could be due to the reevaporation of volatile silicon hydrides formed on the substrate surface. The dependence of the deposition rate on the acceleration voltage was also

TABLE III

DEPOSITION CONDITIONS

Acceleration voltage, V_a	0–8 kV
Electron current for ionization, I_e	100 mA
Acceleration voltage of electron beam for ionization, V_e	150 V
Hydrogen gas pressure, P_{H_2}	$0-3 \times 10^{-4}$ Torr
Doping gas concentration in hydrogen gas	0–5000 vppm[a]
Substrate temperature, T_s	RT–300°C

[a] vppm is volume parts per million.

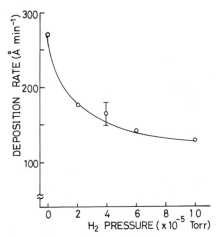

FIG. 13. Dependence of deposition rate on hydrogen gas pressure (one crucible with single nozzle), $V_a = 0$ kV $T_s = 220\,^\circ$C.

measured (Yamada *et al.*, 1980b). The energy per atom inside the accelerated ionized clusters is less than a few electron volts, so that sputtering by the ions should be negligibly small. The decrease in the deposition rate should be due to an increased chemical reaction of silicon with hydrogen; increasing the acceleration voltage increases the adatom migration on the substrate surface and enhances its chemical reactivity. Figure 14 shows the dependence of the optical band gap of the films on the hydrogen gas pressure in the

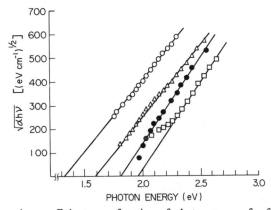

FIG. 14. Absorption coefficient as a function of photon energy for films deposited at different hydrogen gas pressures, $V_a = 3$ kV, $T_s = 220\,^\circ$C. Hydrogen pressure (Torr): $\bigcirc$, 0; $\triangle$, 3×10^{-5}; $\bullet$, 1×10^{-4}; $\square$, 3×10^{-4}.

chamber. The films were deposited at 3 kV on glass substrates heated at 220°C. Sufficient change in the optical band gap of 1.3–1.9 eV could be obtained by changing the hydrogen gas pressure in a range from $0-3 \times 10^{-4}$ Torr.

In the ICB deposition of a-Si:H films the composition of the particles impinging on the substrate surface can be easily estimated because no plasma is present in the deposition chamber. Figure 15 shows the relative numbers of particles impinging on the substrate surface. Since the background gas pressure before introducing the hydrogen gas was 5×10^{-7} Torr, the main particles impinging on the substrate surface are ionized and neutral silicon clusters from the ion source, and the mixed hydrogen gas and doping gases that are introduced into the chamber through the leak valves. The constituent parts of these gases are atomic and molecular hydrogen and their ions, because some of the hydrogen molecules are ionized and some are dissociated inside the ionizing section. Under typical deposition conditions, the number of impinging silicon atoms was in the order of $10^{15}-10^{16}$ atoms cm^{-2} sec^{-1}, as calculated from the measured silicon ion-current density under the assumption that the average size of the cluster was 10^3 atoms per cluster and that 10% of the total beam flux was ionized. The impinging rate of hydrogen ions was calculated to be 10^{13} ions cm^{-2} sec^{-1} from the ion current to the substrate. The ratio of the hydrogen atoms to the hydrogen molecules was estimated from the change of H and H_2 peaks in a mass spectrum when the electrical input power into the source was varied. A 10% increase in hydrogen atoms was observed when the input power was increased above the typical deposition condition. Since the rate of impinging

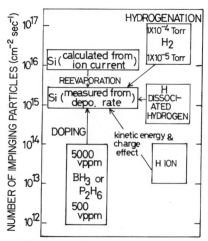

FIG. 15. Kinds of particles impinging on the substrate surface and estimated number of impinging particles.

hydrogen molecules is on the order of 10^{16} molecules cm^{-2} sec^{-1} under operating conditions, approximately 10^{15} atoms cm^2 sec of dissociated hydrogen atoms impinge on the substrate surface. Thus the rate of impinging hydrogen ions is three orders of magnitude smaller than that of molecular hydrogen. It is not clear yet which state of hydrogen (atomic, molecular, or ionic) is dominantly involved in the hydrogenation process. But it seems natural to consider that hydrogen ions could have a considerable influence in providing uniform hydrogenation. It is certain that the reaction takes place on the substrate surface because the mean free path of the particles is much larger than the distance between the substrate and the crucible, and the temperature of silicon in the crucible is too high to contain hydrogen.

VI. Effect of the Acceleration Voltage

Figure 16 shows the SEM images of the films together with a-Si films obtained by conventional vacuum evaporation (Yamada *et al.*, 1980c). The surface of the a-Si films obtained by the ICB is very smooth without a bump of a size exceeding 10 Å. This characteristic seems to be due to the increased

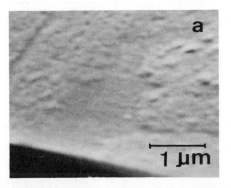

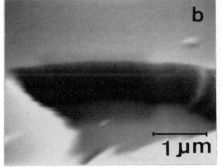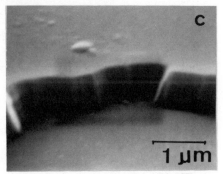

FIG. 16. SEM images of amorphous silicon films: (a) Vacuum deposition; (b) ICB at acceleration voltage $V_a = 0$ kV; (c) ICB at 3 kV.

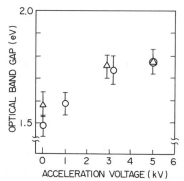

FIG. 17. Optical band gap as a function of the acceleration voltage, $T_s = 220°C$, $P_{H_2} = 1 \times 10^{-4}$ Torr. O, as deposited; △, annealed at 400°C.

nucleation density and adatom migration effect. The adhesion strength of the film on the substrate surface could be increased by increasing the acceleration voltage. Numerical values for the adhesion strength in particular substrate film combinations are given elsewhere (Yamada *et al.*, 1978).

Figure 17 shows the change of the optical band gap of films deposited at different acceleration voltages. The optical band gap increases with increasing acceleration voltage. The voltage show that the hydrogenation of the films can be enhanced at higher acceleration voltage. Annealing characteristics were measured with films deposited at different acceleration voltages. The optical band gap of films deposited at higher acceleration voltage does not change after annealing, as shown in the same figure. The annealing was done at 400°C in vacuum for 30 min. This result shows that it is possible to form thermally stable films. The thermal stability of the films was also checked by ion backscattering (Kubota *et al.*, 1980). Figure 18 shows that the hydrogen content of the film is unchanged after heat treatment up to 450°C. This film was deposited at 3 kV onto a 220°C substrate. The hydrogen gas pressure was 5×10^{-5} Torr. Y_a and Y_c are the backscattering yields from a-Si:H and crystalline silicon, respectively.

The infrared absorption of the films deposited at different acceleration voltages was measured. Figure 19 shows the dependence of the IR absorption. The films showed a strong monohydride absorption at 2000 cm^{-1}. This absorption increased with increasing acceleration voltage whereas the weak absorption by the dihidrides at 2100 cm^{-1} decreased. The results show that the film quality can be controlled by changing the acceleration voltage and that the monohydride structure is dominant in films formed by ICB. The uniform hydrogenation at very low hydrogen gas pressure could be explained as follows: Since the migration of silicon adatoms can be enhanced by transfer energy from the kinetic energy given by the acceleration of the

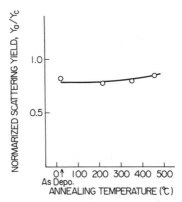

FIG. 18. Change of the hydrogen content of the films after different heat treatments. Yield is related to the change of hydrogen content in the hydrogenated films. Acceleration voltage is 3 kV; H_2 flow is 1×10^{-4} Torr.

ionized clusters, the probability of reaction with the impinging hydrogen atoms increases, resulting in the effective formation of the monohydride. These films are thermally stable. Because dihydride and polysilane are unstable, these species release hydrogen most easily by heating. The thermal stability of the structure can be appreciated from the results shown in Fig. 17, where stable characteristics could be obtained for films deposited at higher acceleration voltages.

The temperature dependences of the dark conductivity of the films deposited at various acceleration voltages are shown in Fig. 20. Films deposited at acceleration voltages of $V_a = 0 - 2$ kV show the typical

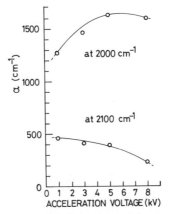

FIG. 19. Absorption coefficient as a function acceleration voltage, $T_s = 220°C$, $P_{H_2} = 1 \times 10^{-4}$ Torr.

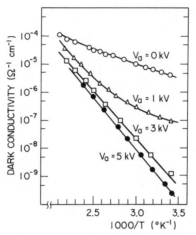

FIG. 20. Temperature dependence of the dark conductivity of the films deposited at different acceleration voltages, $T_s = 220°C$, $P_{H_2} = 1 \times 10^{-4}$ Torr.

hopping conduction between localized states, while the films deposited at $V_a = 3-5$ kV show the extended-state conduction with a well-defined activation energy. The dependences of photoconductivity and dark conductivity on the acceleration voltage are shown in Fig. 21. The films were deposited in hydrogen of $P_{H_2} = 1 \times 10^{-4}$ Torr at a substrate temperature $T_s = 220°C$. The photoconductivity was calculated from the difference

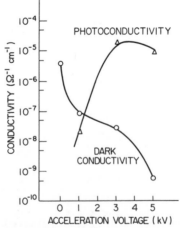

FIG. 21. Dark conductivity and photoconductivity of the films as a function of the acceleration voltage, $T_s = 220°C$, $P_{H_2} = 1 \times 10^{-4}$ Torr.

between the current measured under illumination and the dark current at a constant electric field of 1.7×10^3 V cm^{-1}. The dark conductivity of the films decreased by four orders of magnitude by increasing the acceleration voltage. The photoconductivity approaches 1×10^{-4} (Ω cm)$^{-1}$ with an incident photon flux of 3.19×10^{17} photons (cm^{-2} sec^{-1}) from a He–Ne laster (100 mW cm^{-2}).

The effect of the acceleration voltage on the photo- and dark conductivities may be explained by assuming that the number of hydrogenated dangling bonds in the deposited film increased at higher acceleration voltage. This is attributed to an increased probability of chemical reaction between the depositing silicon and the hydrogen atoms due to the enhanced adatom migration on the substrate surface when the acceleration voltage is increased.

VII. Characteristics of Doped Films

In the ICB deposition the doping materials need not always be supplied by decomposition of gases, as in the case of the glow-discharge method. Therefore, several doping methods can be used, such as simultaneous evaporation of doping elements, deposition from an already doped source, deposition in a doping gas, ion implantation of the doping material, and so on, because the film is formed in a relatively low gas pressure region and the evaporating material is transported as a beam of clusters. In this experiment, doped films have been formed by deposition in a mixture of hydrogen and phosphine

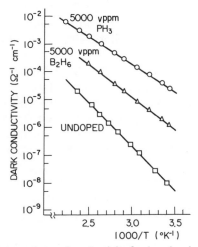

FIG. 22. Temperature dependence of conductivity for doped and undoped films, $V_a = 3$ kV, $T_s = 300°C$, $P_{H_2} = 1 \times 10^{-4}$ Torr.

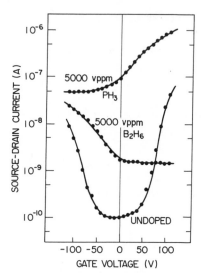

FIG. 23. I–V characteristics of MOS devices with doped and undoped films. Depositions were made under the same conditions as those in Fig. 22.

(PH$_3$) or diborane (B$_2$H$_6$), the dopants being in a concentration of about 5000 vppm.

Figure 22 compares the dark conductivity of phosphorus- and boron-doped films with an undoped one. The films were prepared at the acceleration voltage $V_a = 3$ kV and substrate temperature $T_s = 300°$C in the mixed gas pressure $P_{H_2} = 1 \times 10^{-4}$ Torr. The activation energy of these films changes from 0.64 eV for the undoped film to 0.48 eV for the boron-doped one and to 0.41 eV for the phosphorus-doped one. Field effect measurements were made to confirm electrical-type conversion. Figure 23 shows the source-to-drain current and the gate voltage characteristics of the MOS devices with the films deposited in different type of mixed gases. In the phosphorus-doped film, for which the current remains substantially flat with increasing negative gate voltage, the current is predominantly carried by electrons. In the sample doped with boron, hole current is predominant. From these results, it is clear that n- and p-type films could be formed. For the undoped film the curve shows a slightly n-type conduction.

Figure 24 shows the conductivities and the activation energies of the films deposited at different mixed gas concentrations. The conductivity and the activation energy can be controlled by changing the doping gas concentration.

A-Si: H p–i–n diodes were fabricated on stainless steel, tantalum, molybdenum, and ITO/glass substrates. The diodes were made in a single chamber

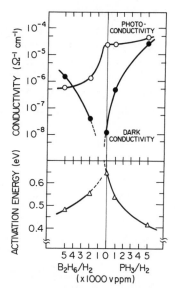

FIG. 24. Conductivities and activation energies as a function of the doping gas concentration $V_a = 3$ kV, $T_s = 300°C$, $P_{H_2} = 1 \times 10^{-4}$ Torr.

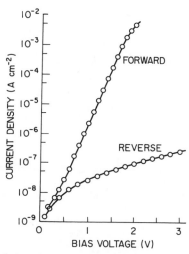

FIG. 25. I–V characteristics of a $p–i–n$ diode deposited on Ta substrate at $V_a = 3$ kV, $T_s = 300°C$, $P_{H_2} = 1 \times 10^{-4}$ Torr. Doping gas concentration in hydrogen was 5000 vppm (volume parts per million).

by changing the flow of doping gas (diborane or phosphine) by adjusting the leak valves. Impurities in the films were analyzed by SIMS (secondary ion mass analyzer). No trace of phosphorus or boron in the *i* layer could be observed. Figure 25 shows the variation of the dark *I-V* characteristic of the films deposited on a tantalum substrate. The layer thicknesses of *p*, *i*, and *n* layers are 250, 3500, and 250 Å, respectively. Although optimization of the deposition conditions and structural design has not been done yet, the diode shows good rectification characteristics. By optimizing the film deposition conditions and thickness, the device characteristics could be improved (Yamada *et al.*, 1983).

VIII. Conclusions

In the ICB deposition of amorphous silicon films, it was found that the energy of the accelerated ionized clusters and the ionic charge contributed to the formation of hydrogenated film in relatively low hydrogen pressure, due to the enhanced chemical reactivity of silicon with hydrogen. In the actual deposition, the hydrogenated film could be formed in a hydrogen gas pressure of $10^{-5}-10^{-4}$ Torr due to the effects of the kinetic energy and the ionic charge of the ionized particles. The films show predominantly mono-hydride structure and thermally stable characteristics. Only about 5000 vppm of doping gas in hydrogen was enough to change the conductivity type of the films. In this method it is not necessary to introduce carrier gases to sustain a plasma so that gas precipitation troubles are avoided. Moreover, undesired gas and impurity incorporation can be eliminated because the deposition is carried out at relatively low gas pressure. Films having a smooth surface, good adhesion, and thermally stable characteristics could be formed. Production scale of the equipment is possible by applying the multiple-nozzle or multiple-crucible system. These experiments show that ICB deposition can be used to form thin a-Si:H films for electron device applications such as solar cell, FET, and so on.

ACKNOWLEDGMENTS

The author gratefully acknowledges T. Takagi, Director of the Ion Beam Engineering Experimental Laboratory, Kyoto University, for supporting the ICB experiment. Thanks are also due to H. Usui, M. Horie, I. Nagai, and S. Ishiyama of Takagi's laboratory, Kyoto University, for their exellent experiments.

REFERENCES

Babaev, V. O., Bykov, Ju. V., and Guseva, M. B. (1976). *Thin Solid Films* **38**, 1-8.
Hagena, O. F. (1974). *Phys. Fluids* **17**, 894-896.

Hilschfelder, J. O., Curtiss, C. F., and Bird, R. B. (1964). "Molecular Theory of Gases and Liquids," p. 336–391. Wiley, New York.

Kristensen, W. D., Jensen, E. J., and Cotterill, R. M. J. (1974). *J. Chem. Phys.* **60**, 4161–4168.

Kubota, K., Imura, T., Iwami, M., Hiraki, A., Satou, M., Fujimoto, F., Hamakawa, Y., Minomura, S., and Tanaka, K. (1980). *Nucl. Instrum. Methods* **168**, 211–215.

Lee, J. K., Barker, J. A., and Abraham, F. F. (1973). *J. Chem. Phys.* **58**, 3166–3180.

Neugebauer, C. A. (1970). *In* "Handbook of Thin Film Technology" (L. I. Maissel and R. Glang, eds.), pp. 8-3–8-44. McGraw-Hill, New York.

Stein, G. D. (1979). *Phys. Teacher* **17**(8), 503–512.

Takagi, T. (1982a). *Thin Solid Films* **92**, 1–17.

Takagi, T. (1982b). *Proc. N. C. State Univ. Conf. Ceram. Sci. 19th, 1982* (to be published).

Takagi, T. (1982c). Preprint of *Ion Assisted Surf. Treat., Tech. Processes, Univ. Warwick, Coventry, England, 1982* pp. 1.1–1.8.

Takagi, T., Yamada, I., Kunori, M., and Kobiyama, S. (1972). *Proc. Int. Conf. Ion Sources, 2nd, 1972* pp. 790–795.

Takagi, T., Yamada, I., and Sasaki, A. (1975). *J. Vac. Sci. Technol.* **12**(6), 1128–1134.

Takagi, T., Yamada, I., and Sasaki, A. (1976). *Thin Solid Films* **39**, 207–217.

Takagi, T., Yamada, I., and Sasaki, A. (1978a). *Conf. Ser.—Inst. Phys.* **38**, 142–150.

Takagi, T., Yamada, I., and Sasaki, A. (1978b). *Conf. Ser.—Inst. Phys.* **38**, 229–235.

Takagi, T., Yamada, I., and Takaoka, H. (1981). *Surf. Sci.* **106**, 540–550.

Yamada, I. and Takagi, T. (1981). *Thin Solid Films* **80**, 105–115.

Yamada, I., Matsubara, K., Kodama, M., Ozawa, M., and Takagi, T. (1978). *J. Cryst. Growth* **45**, 326–331.

Yamada, I., Saris, F. W., Takagi, T., Matsubara, K., and Takaoka, H. (1980a). *Jpn. J. Appl. Phys.* **19**, L181–L184.

Yamada, I., Nagai, I., Ishiyama, S., and Takagi, T. (1980b). *Proc. Symp. Ion Sources Ion Appl. Technol., 4th, 1980* pp. 115–116.

Yamada, I., Nagai, I., Ishiyama, S., and Takagi, T. (1980c). *Proc. Symp. Ion Sources Ion Appl. Technol., 4th, 1980* pp. 117–118.

Yamada, I., Stein, G. D., Usui, H., and Takagi, T. (1982a). *Proc. Symp. Ion Sources Ion-Assisted Technol., 6th, 1982* pp. 47–52.

Yamada, I., Takaoka, H., Inokawa, H., Usui, H., Cheng, S. C., and Takagi, T. (1982b). *Thin Solid Films* **92**, 137–146.

Yamada, I., Nagai, I., Horie, M., and Takagi, T. (1983). *J. Appl. Phys.* **54**(3), 1583–1587.

SEMICONDUCTORS AND SEMIMETALS, VOL. 21A

CHAPTER 6

Chemical Vapor Deposition

Masataka Hirose

DEPARTMENT OF ELECTRICAL ENGINEERING
HIROSHIMA UNIVERSITY
HIGASHIHIROSHIMA, JAPAN

I. Introduction

Discharge-produced amorphous silicon contains 15 – 18 at. % of bonded-hydrogen atoms. The presence of hydrogen relaxes the strained bonds in the tetrahedral network and reduces silicon dangling bonds and the corresponding gap states. In contrast, chemically vapor-deposited amorphous silicon is composed of an almost pure silicon network because the thermal decomposition of silane needs a temperature higher than 500°C, at which bonded hydrogen, if it exists on the growing surface, is thermally emitted into the gas phase. Consequently, gap states in CVD a-Si are much more numerous than in discharge-produced a-Si : H.

Electronic properties of CVD amorphous silicon were first reported in 1977 (Hirose *et al.,* 1977) and substitutional doping was achieved despite the high density of defect states (Taniguchi *et al.,* 1978). The CVD a-Si has been utilized as high-temperature photothermal converters (Seraphin, 1976) and as an optical detector operating in the picosecond range (Auston *et al.,* 1980). Recently, it was found that posthydrogenation of CVD a-Si by hydrogen plasma annealing dramatically improves the electronic properties of the thin films (Sol *et al.,* 1980). This technique has been applied to produce photovoltaic cells (Nakashita *et al.,* 1981) and more recently to prepare high-current pin rectifiers (Szydlo *et al.,* 1982). Another new direc-

109

tion in the preparation of hydrogenated CVD a-Si is to lower the deposition temperature down to 300–400°C, using a homogeneous CVD technique (Scott *et al.*, 1981) or a low-temperature CVD of higher silanes (Weinberger *et al.*, 1981). Recently, amorphous hydrogenated silicon has been deposited by photochemical decomposition (photo-CVD) of disilane at a temperature below 300°C (Mishima *et al.*, 1983). The progress in the preparation technique has enabled us to obtain device-quality CVD a-Si films at low temperatures. Extensive studies of CVD a-Si with and without hydrogenation have provided important information on the growth mechanism of amorphous hydrogenated silicon (Scott *et al.*, 1981), have established the minimum concentration of bonded hydrogen necessary for eliminating the spin (Sol *et al.*, 1980), and have revealed the microscopic structure of native defects in amorphous silicon (Hirose, 1981).

The CVD technique is widely used in the electronic industry, and deposition apparatus suitable for mass production is available. Hence this technique is well suited for fabricating large-area amorphous silicon devices in scaled-up reactors.

II. CVD at Low Temperature

The thermal decomposition of monosilane at a temperature around 1050°C is currently used for growing epitaxial layers on silicon substrates. Also, doped polycrystalline silicon has been utilized as interconnections or self-aligned gate electrodes in integrated circuits. In the case of polycrystalline films, the deposition temperature is usually 650–700°C and subsequently the films are annealed at ~ 1050°C, so that the as-grown films are either amorphous or fine-grained polycrystalline because the transition from the amorphous to the crystalline phase occurs at a temperature of 675–690°C, whereupon an abrupt change in the refractive index takes place as shown in Fig. 1a and b (Hirose *et al.*, 1977). Despite the presence of many extensive studies on structural and transport properties of high-temperature CVD silicon, there had been very little work on the basic properties of low-temperature CVD films until the technological importance of discharge-produced amorphous silicon was demonstrated by the success of substitutional doping (Spear and LeComber, 1976).

1. DEPOSITION TECHNIQUE AND CHEMICAL PROCESS

Amorphous CVD silicon is prepared by the thermal decomposition of silane at a temperature between 500 and 650°C using a deposition system schematically illustrated in Fig. 2. Silane gas diluted with 99% H_2 (or Ar) is admitted into a quartz tube reactor at a flow rate of ~ 1 liter min^{-1} together

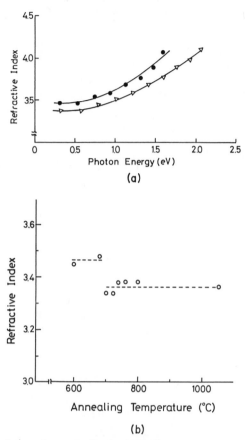

FIG. 1. (a) Spectral dependence of refractive index for an as-grown amorphous silicon film and for a polycrystalline silicon film. ●, deposited at 650°C; ▽, annealed at 1050°C. (From Hirose *et al.,* 1979.) (b) Refractive index as a function of annealing temperature for a CVD silicon specimen grown at 600°C. (From Hirose *et al.,* 1979.)

with nitrogen carrier gas ($\sim$ 1 liter min^{-1}) in the case of atmospheric pressure CVD, while a few percent of silane in H$_2$ or Ar is fed to the reactor in low-pressure CVD. The growth rate of undoped a-Si has an activation energy of 1.5 eV (34.6 kcal mol^{-1}), as shown in Fig. 3 (Hirose, 1981). The activation energy is significantly increased with phosphorus doping and dramatically reduced by boron doping, possibly because the growth kinetics is appreciably influenced by the catalytic effect of doping atoms sitting on the growing surface (Yasuda *et al.,* 1974; Farrow and Filby, 1971).

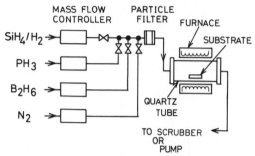

FIG. 2. Schematic view of apparatus for preparing amorphous CVD silicon.

The reaction steps in the gas phase and on the substrate surface are inferred as follows (Ban *et al.*, 1973):

$$SiH_4(g) \rightarrow SiH_2(g) + H_2(g) \tag{1}$$

$$SiH_2(g) + Si(s) \rightarrow 2Si-H(a) \tag{2}$$

$$2Si\text{-}H(a) \rightarrow 2Si(s) + H_2(g) \tag{3}$$

Here (g), (s), and (a) denote the gas, solid, and adsorbed molecules, respectively. When the reaction of Eq. (3) is slowest and hence a rate-limiting

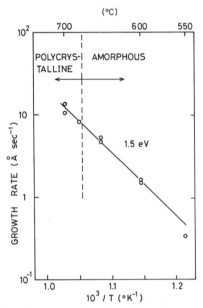

FIG. 3. Growth rate of CVD a-Si as a function of deposition temperature. (From Hirose, 1981.)

process of the deposition at low temperatures, the growth rate is given by

$$\frac{dr(\text{Si})}{dt} = K_1[P_{\text{SiH}_4}]^{1/2}[P_{\text{H}_2}]^{-1/2} - K_2[P_{\text{H}_2}], \qquad (4)$$

where $dr(\text{Si})/dt$ represents the growth rate, $[P_x]$ is the partial pressure of the gas x, and K_1 and K_2 are proportionallity constants. The second term in the right-hand side is negligible at temperatures around 600°C and therefore the growth rate is proportional to the first term in the right-hand side of Eq. (4), as experimentally demonstrated by Ban *et al.* The activation energy of 34.6 kcal mol^{-1} for undoped films is in good agreement with a value reported by Bryant (1979), who also pointed out that the rate control was provided by the heterogeneous reactions.

In the conventional CVD of silane, low-cost substrates such as glass or stainless steel cannot be used on account of the high-temperature process. In this respect, homogeneous CVD or the use of disilane are possible solutions for lowering the deposition temperature (see the later sections). Another new attempt to grow hydrogenated silicon is to employ photochemical decomposition of monosilane with mercury photosensitization (Sarkozy, 1981; Ito *et al.*, 1982). The decomposition mechanism of SiH$_4$ is believed to involve a collisional energy transfer from the UV-excited Hg* (^{3}P$_1$) state to the reacting species (Niki and Mains, 1964). Mercury incorporation into the deposited thin films appears to be negligible. Nevertheless, it is necessary to develop an appropriate method for achieving direct photochemical decomposition without mercury photosensitization, because residual Hg contamination in the reaction chamber should be eliminated. It was recently found that disilane is decomposed by direct photoexcitation of UV light from a Hg resonance lamp ($\lambda = 2537$ Å, 1849 Å), although no decomposition happens in monosilane (Mishima *et al.*, 1983). Figure 4 is a schematic diagram of a photo-CVD system. A low-pressure mercury lamp (110 W) as a UV-radiation source for the direct excitation of Si$_2$H$_6$ gas was placed just above a horizontal quartz reactor. Photochemical vapor deposition was carried out at atmospheric pressure, and 1% Si$_2$H$_6$ in He was admitted into the quartz

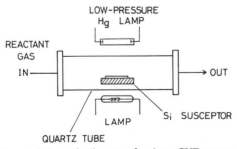

FIG. 4. Schematic diagram of a photo-CVD system.

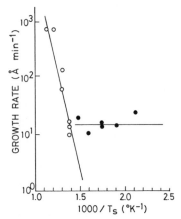

FIG. 5. Growth rate of photo-CVD and thermal CVD films produced from disilane versus substrate temperature. ○, thermal; ●, photochemical. (From Mishima *et al.*, 1983.)

tube at a flow rate of 400 sccm, together with a nitrogen carrier gas at a flow rate of 500 sccm. The substrate was placed on a silicon susceptor heated by two halogen lamps situated below the reactor.

Plots of the deposition rate versus reciprocal substrate temperature for silicon thin films produced by thermal CVD and by direct photo-CVD of Si_2H_6 are compared in Fig. 5. Extrapolation of the growth rate for thermal CVD activated with an energy of 1.6 eV at temperatures below 300°C provides a rate of less than 0.1 Å min^{-1}, whereas the direct photo-CVD provides a constant growth rate of ~ 15 Å min^{-1} even below 300°C for an incident UV power density of 0.08 W cm^{-2}. More than two orders of magnitude increase in the growth rate by UV excitation were obtained and the deposition rate enhancement by an increase of UV incident power indicates that the direct photochemical reaction of Si_2H_6 proceeds in the gas phase and/or on the surface. The observed rate of ~ 15 Å min^{-1} will be improved to 30 Å min^{-1}, which is a typical value obtained for the mercury-photosensitized decomposition of SiH_4 using 2537-Å light.

2. MATERIALS PROPERTIES

The ideal network of amorphous silicon is considered to possess values of refractive index and density close to those of the crystal. The infrared refractive index of CVD a-Si is only 3% greater than that of the polycrystalline silicon, as shown in Fig. 1a, and the film density is 97% of the crystalline value. No infrared absorption due to bonded hydrogen is observed in CVD a-Si, and the hydrogen content might be far below 0.5 at. % as estimated by a detection limit of the IR spectrometer. Note that the dangling-bond density

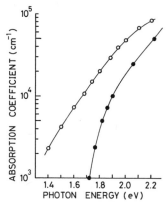

FIG. 6. Optical absorption spectra for CVD a-Si and discharge-produced a-Si : H. ○, CVD; ●, plasma deposited. (From Hirose, 1981.)

measured by ESR tends to decrease from 1×10^{19} to 2.5×10^{18} cm^{-3} when the deposition temperature is lowered from 650 to 500°C, possibly due to a minute incorporation of hydrogen into the network during a-Si growth. Thus it is likely that CVD a-Si is composed of a nearly ideal silicon network without hydrogen incorporation and without a density deficit. Figure 6 represents the optical absorption coefficient α measured as a function of photon energy for CVD a-Si. The optical band gap determined by a $(\alpha h v)^{1/2}$ versus $h v$ plot is 1.45 eV, which is in agreement with a value obtained by extrapolating the hydrogen concentration dependence of the optical gap for discharge-produced a-Si : H to zero concentration (Matsuda *et al.*, 1981).

3. SUBSTITUTIONAL DOPING AND ESR

The conductivity of CVD a-Si can be controlled over a wide range by phosphorus or boron doping as shown in Fig. 7. The figure also shows the doping dependence of the spin density. The g value of 2.0054 and the linewidth of 5.6 ~ 6.7 G are almost independent of the dopant element and of its concentration. The one order of magnitude decrease in n-type conductivity with phosphorus doping at ratios below 4×10^{-4} is attributable to a decrease in the density of gap states because room temperature conductivity is dominated by the hopping conduction through states near the Fermi level (Taniguchi *et al.*, 1978). Indeed, the measured gap state density near the Fermi level in a-Si is reduced by one order of magnitude with a phosphorus doping of 4×10^{-4} (Hirose *et al.*, 1979). Above a doping ratio of 4×10^{-4}, the spin density is greatly reduced and the conductivity increases as a result of the upward motion of the Fermi level. The position of the Fermi level is determined by the conductivity activation energy at high temperatures and

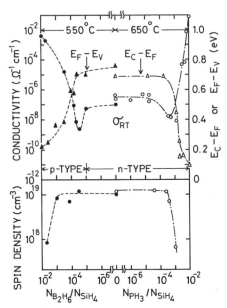

FIG. 7. Room temperature conductivity σ_{RT}, position of the Fermi level determined from conductivity activation energy, and spin density plotted as a function of doping ratio. (From Hirose, 1981.)

by the temperature coefficient of the mobility gap. It should be noted that the spin density starts to decrease at a ratio of 4×10^{-4}, above which the density of phosphorus atoms exceeds a spin density of 1.5×10^{19} cm^{-3} found in undoped films. A smooth movement of the Fermi level at doping ratios above 4×10^{-4} indicates that a significant part of the incorporated phosphorus atoms acts as donors. However, the rest of the dopant atoms serve to eliminate ESR centers by forming an impurity–defect complex (Hirose *et al.,* 1980).

In the case of boron doping, again the conductivity decreases with doping up to a ratio of 3×10^{-5}, mainly because the Fermi level shifts toward the valence band and p-type conductivity appears at doping ratios above 1×10^{-5}, and the hopping conduction component, which dominates room temperature conductivity, is less influenced with boron doping below 3×10^{-5}. Note that the spin density remains unchanged by doping up to 1×10^{-3} in contrast to the case of phosphorus doping. The density of spins begins to decrease when the Fermi level reaches $E_v + 0.37$ eV and is reduced by one order of magnitude at a boron doping of 5×10^{-3}, where upon $E_F = E_v + 0.30$ eV. If the ESR centers are assumed to be located in the narrow energy range ΔE around $E_v + 0.37$ to $E_v + 0.30$ eV and boron acceptor levels are deeper than this energy range, the Fermi level movement across the energy range ΔE results in the removal of unpaired electrons. The value of ΔE is estimated to be about 0.1 eV using the relationship $N(E_T)$ $\Delta E \simeq N_s$, where $N(E_T)$ is the measured gap state density at the energy E_T and N_s is the measured spin density. This estimated value of ΔE is similar to that obtained for a specimen doped with 3.7×10^{-3} (Hirose, 1981) when the gap state distribution was measured.

III. CVD with Posthydrogenation

From the viewpoint of defect passivation and bond relaxation in plasma-deposited a-Si:H matrix, the incorporation of 18 at. % of hydrogen appears to be too high. Chemically vapor-deposited a-Si contains no detectable bonded hydrogen and a small amount of hydrogen could be introduced into the film either by hydrogen plasma annealing (Sol et al., 1980) or by H$^+$ ion implantation (Suzuki et al., 1980). All the spins in a-Si could, in principle, be saturated by ~0.1 at. % of hydrogen. Indeed, Sol et al. demonstrated that if CVD a-Si is annealed in a dc or microwave glow discharge of hydrogen at 400°C, the ESR signal is eliminated at a hydrogen concentration of the same order as the dangling-bond density.

4. HYDROGEN DEPTH PROFILES AND SPIN ELIMINATION

Hydrogen penetration into CVD a-Si is basically controlled by a diffusion process with a diffusion coefficient of 5.8×10^{-14} cm^2 sec^{-1} at 400°C (Sol et al., 1980). Typical in-depth profiles of samples annealed in a deuterium plasma have been determined by SIMS (secondary ion mass spectrometry)

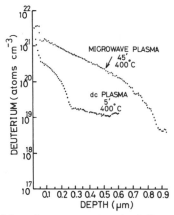

FIG. 8. Concentration of deuterium atoms versus depth for a dc plasma and a microwave plasma at $T = 400$°C. (From Magarino et al., 1982.)

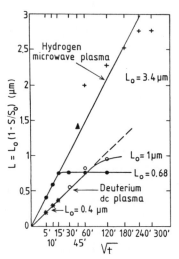

FIG. 9. Length passivated by hydrogenation versus the square root of the treatment duration at $T = 400°C$ and fixed conditions of treatment. Samples with $L_0 > 1$ μm are rather inhomogeneous in depth and the results show a great dispersion. *, Deuterium dc plasma; ○, deuterium dc plasma; ●, hydrogen microwave plasma; +, hydrogen microwave plasma; ▲, deuterium microwave plasma. (From Magarino et al., 1982.)

as shown in Fig. 8. A peak in the deuterium concentration appears near the surface when the plasma is maintained during cooldown (Magarino et al., 1982). This strongly hydrogenated surface region should have electronic properties different from those of the bulk region, so that the observed properties of undoped samples are sometimes influenced by the surface layer.

The principal effect of posthydrogenation is the elimination of the dangling-bond ESR signal. The hydrogen penetration depth is well correlated with the spin-elimination depth $L = L_0(1 - S/S_0)$, where L_0 is the film thickness, S_0 is the initial spin density, and S is the spin signal after the plasma annealing time t. As shown in Fig. 9, the linear dependence of L versus $\sqrt{t}$ is consistent with a spin-elimination front advancing through a diffusion process.

A similar result is obtained by rf plasma annealing; the spin density measured as a function of film thickness provides the effective spin-elimination depth as shown in Fig. 10 (Hirose, 1981). The figure indicates that a very low amount of diffused hydrogen, less than 0.5 at. %, can eliminate the spins. Following the saturation of dangling bonds, the AM1 photoconductivity exceeds 1×10^{-3} Ω^{-1} cm^{-1} and the photoluminescence intensity of a main peak centered at 1.23 eV becomes about 50 times as large as that of an as-deposited specimen. The peak energy of the PL spectra is apparently low

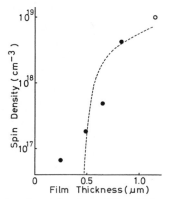

FIG. 10. Spin density measured as a function of film thickness for CVD a-Si annealed in hydrogen plasma at 350°C and 40 W for 1 hr. ○, as deposited; ●, hydrogenated; ---, calculated. (From Hirose, 1981.)

compared to that of discharge-produced a-Si : H, because the optical band-gap of hydrogenated CVD a-Si is 1.55 eV, which is by about 0.2 eV smaller than that of a-Si : H. The in-depth profile of the PL intensity was determined during a step-by-step etching of a silicon layer. A rapid decrease in intensity is found because of the decreasing concentration of hydrogen diffusing into the bulk.

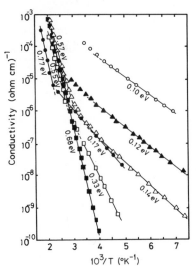

FIG. 11. Temperature dependence of dc conductivity for hydrogen-implanted undoped CVD a-Si annealed at various temperatures. ●, as grown; ○, as implanted. ▲, 200°C anneal; △, 300°C anneal; ■, 400°C anneal; □, 500°C anneal. (From Suzuki et al., 1980.)

The other hydrogenation technique employs H⁺ ion implantation. In this case, implanted hydrogen profile follows the LSS theory (Lindard *et al.,* 1963), and the mean penetration depth depends on the ion acceleration energy (~ 0.3 μm at 20 keV for H⁺ ions). When $0.6 \sim 0.7$-μm-thick films are irradiated with a dose of 5×10^{16} cm⁻² at 20 keV, the average hydrogen content is estimated to be ~ 1.5 at. % (Suzuki *et al.,* 1980). The infrared absorption due to the Si–H stretching mode at 2000 cm⁻¹ is observable even in as-implanted samples and increases upon annealing at 400°C. Postimplantation annealing at 400°C removes most implantation-induced defects as seen in Fig. 11, where the hopping conduction through defect states is eliminated by the 400°C annealing.

5. REACTIVATION OF DOPANTS

Some of the doping atoms in as-grown CVD films are electrically inactive because of electronic compensation by defects; the other doping atoms saturate native defects by forming dopant–defect (possibly multiple divacancy) complexes, in which the doping atom might be threefold coordinated (Hirose, 1981). Posthydrogenation eliminates most of the defects in the network, thus improving dramatically the doping efficiency as can be seen by comparing the results shown in Fig. 12 with those of Fig. 7. Since the dopant concentration in CVD a-Si coincides with the doping ratio in the gas phase (Mishima *et al.,* 1980) and since a part of dopant atoms compensates defect states, the activation process of phosphorus or boron atoms could be interpreted in terms of a saturation of the dangling bonds by diffusing hydrogen. Most of the threefold dopant atom–divacancy complexes might be electrically activated by terminating the dangling bonds in the complex

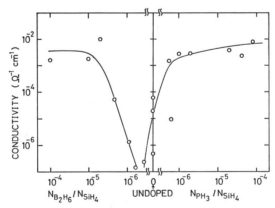

FIG. 12. Room temperature conductivity of posthydrogenated CVD a-Si plotted against doping ratio. (From Hirose, 1981.)

defects. Note that the incorporation of less than 0.5 at. % of hydrogen results in the very efficient activation of dopant atoms. Also, both the dark conductivity and photoconductivity for weakly hydrogenated CVD a-Si exhibit no deterioration by Ar^+ ion laser irradiation of 200 mW cm^{-2} for 2 hr, indicating the absence of electronic and/or structural instabilities.

6. FUTURE PROSPECTS

For the use of low-cost substrates, the deposition temperature of CVD a-Si has to be lowered. A decrease in the growth temperature below 400°C has recently been achieved by utilizing a higher silane as a reactant gas. Using disilane, a photovoltaic cell with a conversion efficiency of about 4% has been realized (Dalal, 1982). Another attractive technique to reduce the growth temperature is to cool the substrate holder in the conventional CVD reactor (Scott *et al.*, 1981). In this case the homogeneous decomposition of SiH_4 produces the film precursor, SiH_2, as a result of the reaction of Eq. (1) and the release rate of hydrogen from the substrate surface appears to be remarkably suppressed. Amorphous silicon thus prepared contains bonded hydrogen in increasing quantity when the substrate temperature is lowered. This kind of approach will provide new and important knowledge of the surface reaction of various chemical species and of the growth mechanism of amorphous hydrogenated silicon. The direct photo-CVD technique is also a promising method to obtain high-quality a-Si : H at low temperature (Mishima *et al.*, 1983). The dark conductivity and photoconductivity are typically 10^{-10} ohm^{-1} cm^{-1} and 4×10^{-4} Ω^{-1} cm^{-1} (AM1), respectively. Very recently, a highly conductive SiGeB alloy has been prepared by low-pressure CVD (Murase *et al.*, 1982). The film is chemically inert and thermally stable and possible to use as a universal contact in semiconductor devices. Extensive efforts to realize low-temperature CVD and to synthesize a new class of hydrogenated amorphous silicon will lead to further progress in CVD techniques that are basically suited for mass production.

REFERENCES

Auston, D. H., Lavallard, P., and Kaplan, D. (1980). *Appl. Phys. Lett.* **36**, 66.
Ban, Y., Tsuchikawa, H., and Maesa, K. (1973). *Semicond. Silicon, Pap. Int. Symp. Silicon Mater. Sci. Technol., 2nd, 1973* p. 292.
Bryant, W. A. (1979). *Thin Solid Films* **60**, 19.
Dalal, V. (1982). *Proc. U.S.-Jpn. Jt. Semin. Technol. Appl. Tetrahedral Amorphous Solids, 1982.*
Farrow, R. F. C., and Filby, J. D. (1971). *J. Electrochem. Soc.* **118**, 149.
Hirose, M. (1981). *J. Phys., Suppl.* **10**, C4-705.
Hirose, M., Taniguchi, M., and Osaka, Y. (1977). *Amorphous Liq. Semicond. Proc. Int. Conf., 7th, 1977* p. 352.
Hirose, M., Taniguchi, M., and Osaka, Y. (1979). *J. Appl. Phys.* **50**, 377.

Hirose, M., Taniguchi, M., Nakashita, T., Osaka, Y., Suzuki, T., Hasegawa, S., and Shimizu, T. (1980). *J. Non-Cryst. Solids* **35–36,** 297.

Ito, H., Hatanaka, M., Mizuguchi, K., Miyake, K., and Abe, H. (1982). *Proc. Symp. Dry Process, 4th, 1982* p. 100.

Lindhard, J. Scharf, M., and Schiøtt, H. E. (1963). *Mat. Fys. Medd.—K. Dan. Vidensk. Selsk.* **33,** 1.

Magarino, J., Kaplan, D., and Friederich, A. (1982). *Philos. Mag. [Part] B* **45,** 285.

Matsuda, A., Matsumura, M., Yamasaki, S., Yamamoto, H., Imura, T., Okushi, H., Iizima, S., and Tanaka, K. (1981). *Jpn. J. Appl. Phys.* **20,** L183.

Mishima, Y., Hirose, M., and Osaka, Y. (1980). *J. Appl. Phys.* **51,** 1157.

Mishima, Y., Hirose, M., Osaka, Y., Nagamine, K., Ashida, Y., Kitagawa, N., and Isogaya, K. (1983). *Jpn. J. Appl. Phys.* **22,** L46.

Murase, K., Amemiya, Y., and Mizushima, Y. (1982). *Jpn. J. Appl. Phys.* **21,** 1559.

Nakashita, T., Hirose, M., and Osaka, Y. (1981). *Jpn. J. Appl. Phys.* **20,** *Suppl.* **20-2,** 201.

Niki, H., and Mains, G. J. (1964). *J. Phys. Chem.* **68,** 304.

Sarkozy, R. F. (1981). *Tech. Dig.—Symp. VLSI Techol. 1981* p. 68.

Scott, B. A., Plecenik, R. M., and Simonyi, E. E. (1981). *Appl. Phys. Lett.* **39,** 73.

Seraphin, B. O. (1976). *Thin Solid Films* **39,** 87.

Sol, N., Kaplan, D., Dieumegard, D., and Dubreuil, D. (1980). *J. Non-Cryst. Solids* **35–36,** 291.

Spear, W. E., and LeComber, P. G. (1976). *Philos. Mag.* [8] **33,** 935.

Suzuki, T., Hirose, M., and Osaka, Y. (1980). *Jpn. J. Appl. Phys. Suppl.* **19-2,** 91.

Szydlo, N., Chartier, E., Proust, N., Magarino, J., and Kaplam, D. (1982). *Appl. Phys. Lett.* **40,** 988.

Taniguchi, M., Hirose, M., and Osaka, Y. (1978). *J. Cryst. Growth.* **45,** 126.

Weinberger, B. R., Akhtar, M., Gau, S. C., and Kiss, Z. (1981). *In* "Tetrahedrally Bonded Amorphous Semiconductors" (R. A. Streer and J. C. Knights, eds.). Amer. Inst. Phys, New York.

Yasuda, Y., Hirabayashi, K., and Moriya, T. (1974). *J. Jpn. Soc. Appl. Phys.* **43,** 400.

SEMICONDUCTORS AND SEMIMETALS, VOL. 21A

CHAPTER 7

Homogeneous Chemical Vapor Deposition

Bruce A. Scott

IBM THOMAS J. WATSON RESEARCH CENTER
YORKTOWN HEIGHTS, NEW YORK

I. Introduction

When amorphous silicon (a-Si) is prepared by evaporation or sputtering, the resulting material contains a large density of dangling bonds and other types of defects. These defect states lie within the mobility gap, pinning the Fermi level and making a-Si impossible to dope with substitutional impurities. If hydrogen is introduced into the growing film by depositing from a plasma discharge in SiH_4 gas, many of the defects are eliminated and the material is vastly improved (Yonezawa, 1981). In part, hydrogen "ties up" the dangling bonds by creating new structural moieties within the network in the form of SiH and SiH_2 groups. So much more hydrogen is incorporated ($\sim 10\%$) compared to the dangling-bond concentration ($\sim 0.1\%$) that a new material has been created: amorphous hydrogenated silicon (a-Si:H). Why so much hydrogen appears to be necessary, how it is incorporated, and the nature of its various electronic and structural roles in a-Si:H are some of the fascinating questions that are ultimately connected to the microscopic details of film growth. It was the search for such mechanistic information that originally led us to devise the homogeneous chemical vapor deposition (HOMOCVD) method for preparing a-Si:H (Scott *et al.,* 1981a).

Like the plasma technique for depositing a-Si:H, HOMOCVD involves the gas phase formation of chemical intermediates that subsequently diffuse to the substrate and react to produce a film. Instead of using plasma

123

excitation, the gas is decomposed thermally. Silane is pumped at pressures ~1 Torr and temperatures above 600°C over a substrate maintained at a much lower temperature. This allows highly reactive SiH₂ species created in the hot gas to deposit a-Si:H at temperatures considerably below those possible in an isothermal system by heterogeneous pyrolysis. Moreover, in contrast to the complex chemical environment of a SiH_4 plasma, which contains many different radical, ionic, and atomic species, there is only one film-forming intermediate in HOMOCVD. This provides us with a powerful approach for unraveling the details of a-Si:H growth chemistry.

In this chapter we review the HOMOCVD method and its use to study the chemistry of a-Si:H film growth. The properties of films are also reviewed, for the similarities and differences between HOMOCVD, plasma, and conventional CVD material have important implications concerning comparative growth environments and suggest new electronic applications for HOMOCVD a-Si:H. Before examining these topics, we first discuss the solid-state chemistry of a-Si:H and its preparation by conventional methods as background for understanding the evolution of the HOMOCVD technique.

II. Deposition and Solid-State Chemistry of a-Si:H

A compositional–structural context for discussing the relationship between preparation method, growth mechanisms, and properties of a-Si:H is set in the form of an oversimplified phase diagram for the silicon–hydrogen system in Fig. 1. The solid curve, based on HOMOCVD film data (Scott *et*

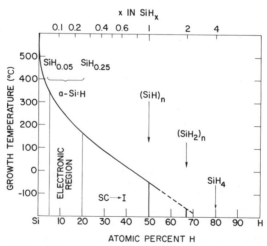

FIG. 1. Temperature–composition diagram for the Si–H system.

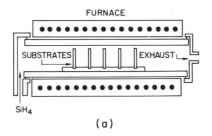

(a)

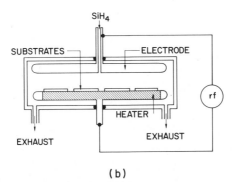

(b)

FIG. 2. (a) Conventional isothermal CVD reactor in which gas and substrate temperatures are equal ($T_g = T_s$) and SiH_4 decomposes on the substrate surface to produce a-Si:H. (b) Capacitively coupled plasma reactor.

al., 1982), illustrates the growth temperature required to achieve a given bulk film composition. It is of course not a true "equilibrium" phase diagram. As will be discussed, comparison of the HOMOCVD composition curve with the results obtained from rf (SiH_4) plasma-deposited films (Milleville *et al.,* 1979; Fritzsche *et al.,* 1979; Knights, 1979) clearly indicate substantial differences; nevertheless, the general compositional trend with growth temperature depicted in Fig. 1 is reflected in data for all film types.

Films close in composition to pure silicon (Fig. 1) are prepared by conventional chemical vapor deposition using SiH_4 gas in an isothermal reactor. A typical system is shown in Fig. 2a. Growth temperatures ($T_g = T_s$) in the vicinity of 500°C or higher must be used because the decomposition of SiH_4 does not proceed at practical rates below this point. For $T_s \sim 500°C$ the hydrogen content C_H is great enough ($\sim 0.5\%$), in principle, to compensate the expected dangling-bond concentration ($N_s \sim 0.1\%$), yet the films exhibit $N_s \sim 10^{19}$ cm^{-3} and poor electronic properties (Hey *et al.,* 1981; Nakashita *et al.,* 1981; Hirose *et al.,* 1977). This is the first clue that no simple connection exists between C_H and N_s. Such CVD films consist of a continuous random network (CRN) of tetrahedral Si(Si$_4$) bonds. The three-dimensionally linked structure is broken every ~ 200 atoms by a Si–H bond. The small hydrogen concentration serves to minimize internal strain by lowering the mean deviation in the average tetrahedral bond angle (Kshirsagar and Lannin, 1982). This important structural ordering effect is discussed by Lannin in Chapter 6 of Volume 21B of this treatise.

Film growth at temperatures below 500°C increases C_H (Fig. 1), enhances structural ordering (Kshirsagar and Lannin, 1982), and lowers the total defect concentration for plasma-deposited material (Yonezawa, 1981; Fritzsche, 1980; Fritzsche et al., 1978). Conventional CVD between 350 and 500°C requires the use of Si_2H_6, Si_3H_8, and higher silanes (Gau et al., 1981a,b), which are thermodynamically and kinetically less stable than SiH_4. However, it is not until film compositions approach the "electronic region" in Fig. 1 that desired devicelike a-Si: H properties are achieved. The upper temperature boundary occurs at $C_H \sim 4-5\%$, where the spin density is in the low 10^{16} cm^{-3} range. C_H is now nearly two orders of magnitude greater than N_s, further illustrating that the two are not simply related. Hydrogen exodiffusion studies (Brodsky et al., 1977a; Oguz and Paesler, 1980) show solid state reactions leading to H_2 loss beginning to take place at 300°C, and annealing experiments (Fritzsche et al., 1978; Biegelsen et al., 1979) indicate a N_s minimum near $T_s = 250$°C. This suggests that the detailed balance is lost, at higher temperatures, between the surface reactions that knit the network together, and H diffusion and annealing effects, occurring simultaneously or following growth, that contribute to the elimination of "errors" (including dangling bonds). For most film deposition methods this loss of coherence in the growth process occurs in the vicinity of 350°C, or $C_H \sim 4-5$ at. %.

Deposition of a-Si: H at substrate temperatures below 350°C has been almost exclusively the domain of the plasma and sputtering methods (Yonezawa, 1981; Fritzsche, 1980), because conventional CVD using higher silane sources (Gau et al., 1981a,b) cannot be achieved with practical growth rates unless very high pressures are used. Substrate temperatures below 350°C are easily attained in HOMOCVD, which is analogous to the plasma method in that highly reactive film-forming intermediates are produced in the gas phase (homogeneously). A capacitively coupled plasma reactor is shown in Fig. 2b. a-Si: H films deposited in such a system at $T_s = 150-350$°C exhibit optimum solid state properties for most applications. In this "electronic region" of Fig. 1, films contain 5–20% H, with $C_H \sim 10-12\%$ typical in actual devices (Hamakawa, 1982). Depending on specific plasma conditions, device-quality films are usually grown in the narrower range: $200 \leq T_s \leq 300$°C. Here they exhibit minimum defect densities (Knights, 1979; Fritzsche, 1980; Fritzsche et al., 1978), including N_s between $10^{15}-10^{16}$ cm^{-3}, and may be doped to high conductivities (Spear and LeComber, 1976).

In the electronic region, films can be considered random alloys, with mostly $Si-H$ bonding at the upper-temperatue solid solution limit, and mixtures of SiH and SiH_2 groups (Knights et al., 1978; Brodsky et al., 1977b) as the low temperature boundary is approached at $C_H \sim 20\%$. The alloy designation is appropriate at the dilute upper limit, but the analogy will break down with increasing C_H as the network becomes progressively less cross-linked (i.e, loses its three-dimensionality). Structurally, this is first manifested in the appearance of isolated SiH_2 groups as the low-temperature solid solution boundary is approached, and $(SiH_2)_n$ clusters (Knights et al., 1978; Brodsky et al., 1977b) as it is traversed at $C_H \sim 20\%$. It should be emphasized that the boundaries designated for the electronic region in Fig. 1 depend very much on preparation method and growth conditions. For example, more hydrogen can be introduced at high substrate temperature by operating a plasma discharge at high power, or a CVD system at high rate and depletion.

As the growth temperature is lowered below $T_s \sim 150°C$, a-Si:H adopts greater polymeric character. The structure becomes progressively more "one-dimensional" as $Si-Si$ bonds are replaced with SiH and SiH_2, and these cluster to create rings and chains of polysilane. Concomitantly, N_s increases, even though C_H is also increasing. The spin density of plasma films grown at $T_s = 25°C$ is nearly as great as unhydrogenated material (Fritzsche et al., 1978), which is surprising because $C_H = 30-40$ at. %. Further, plasma films deposited at $T_s \leq 150°C$ become difficult to dope (Yonezawa, 1981; Spear and LeComber, 1976). This is a consequence of the increasing gap state density, reflected in N_s, and the loss of dimensionality with hydrogen incorporation. As the structure evolves from a dense, rigid CRN comprised of $Si(Si_4)$ units ($C_H = 0$), to one containing almost exclusively $(SiH_2)_n$ chains ($C_H = 67$ at. %), dopant atoms such as P or B, which prefer three-fold coordination, are more easily accommodated in the flexible polymeric network, resulting in the loss of dopant electrical activity.

Thus as C_H moves farther from the electronic region boundary in Fig. 1, a transition in structure occurs, over a relatively large composition range, from a highly cross-linked network composed mostly of Si atoms, SiH units, and isolated SiH_2 groups, to a linear polymer. At $C_H = 50$ at. % the structure can be viewed as containing roughly equal fractions of Si, SiH, and SiH_2, but it is clear from IR spectroscopy (Knights et al., 1978; Brodsky et al., 1977b) that these groups are not distributed completely at random in the network. Films prepared near room temperature contain a considerable quantity of $(SiH_2)_n$ and the structure is sufficiently linearized to have lost its most important properties: photoconductivity, dopability, and narrow band gap. The transition from a three-dimensional amorphous semiconductor to a "one-dimensional" insulator occurs in the composition-temperature space denoted SC → I in Fig. 1. The width of the transition region, and the point of its completion, depends somewhat upon preparation method and experimental conditions. In a-SiH$_x$ prepared by HOMOCVD the transition appears to be complete by $T_s = 25°C$, whereas for plasma material it may lie higher in growth temperature, and could involve a two-phase region (Knights and Lujan, 1979). The nature of the structural and electronic changes in $T-x$ space poses interesting questions in polymer chemistry and percolation theory and is intimately connected with the film growth chemistry.

III. Homogeneous Chemical Vapor Deposition

In addition to hydrogen atoms, the simplest plasma radicals generated by the electron impact of SiH$_4$ are Si, SiH, SiH$_2$, and SiH$_3$ (Turban et al., 1982). Optical emission spectroscopy revealed the presence of Si and SiH in such discharges (Perrin and Delafosse, 1980; Kampas and Griffith, 1981a). a-Si:H film growth from these respective precursors has been argued from diffusion constant (Tachibana et al., 1982) and deposition rate measurements (Taniguchi et al., 1980; Hirose et al., 1981). As discussed in detail elsewhere (Scott et al., 1983a), these radicals, and the ionic species also present, play an important role in the discharge chemistry and secondary reactions of film growth, but SiH$_2$ and SiH$_3$ are favored primary precursors on both energetic and kinetic grounds. A potential energy diagram for the various dissociation reactions is shown in Fig. 3. It is based on known or thermochemically calculated parameters (Walsh, 1981) and published kinetic data (Ring, 1977). Clearly, SiH$_2$ generation would be the most favor-

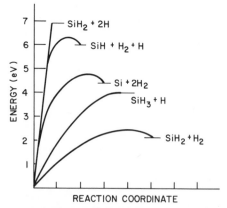

FIG. 3. Potential energy versus (unspecified) reaction coordinate for different silane decomposition channels.

able dissociation pathway in silane decomposition:

$$SiH_4 \rightarrow SiH_2 + H_2. \tag{1}$$

It is now generally accepted that Eq. (1) correctly represents the initial step of homogeneous decomposition in SiH_4 pyrolysis between $600-1100°C$. (Purnell and Walsh, 1966; Newman *et al.*, 1979). From our plasma deposition experiments using higher silanes we hypothesized that SiH_2 could also be the primary plasma deposition intermediate (Scott *et al.*, 1980, 1981b). Films prepared by SiH_4, Si_2H_6, and Si_3H_8 decomposition in low-power inductive rf plasmas under comparable pressure and flow conditions exhibited very similar solid-state properties despite an order of magnitude higher deposition rates from Si_2H_6 and Si_3H_8. This suggested a common deposition chemistry in which plasma dissociation, not subsequent surface reaction, is rate determining. Further, it was shown that hydrogen atoms, generated with much higher concentration in the plasma fragmentation of SiH_4, are not responsible for the lower deposition rate from this gas.

The reactor shown in Fig. 4a was devised to test the idea that SiH_2 is a plausible deposition intermediate (Scott *et al.*, 1981a). By pumping SiH_4 through it at relatively high temperature and pressure, appreciable SiH_2 concentrations are generated in the gas and diffuse to the substrate surface to form a-Si:H film. The crucial difference between the experiment in Fig. 4a and conventional CVD (Fig. 2a) is that $T_s \ll T_g$ in HOMOCVD. This is accomplished by heat-sinking the substrate to a cooled metal block. Since gas temperatures in excess of $600°C$ are used, operation in the isothermal mode of Fig. 2a would result in $C_H < 0.1\%$ (Fig. 1) and poor electronic

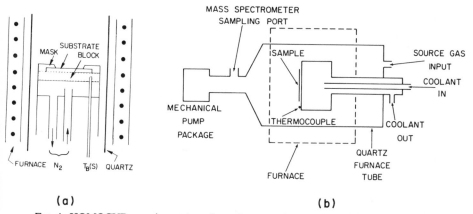

(a)   (b)

FIG. 4. HOMOCVD experimental configuration. A coolant such as N_2 maintains $T_s \ll T_g$. SiH_2 intermediates produced in the SiH_4 gas diffuse to the cold substrate to deposit a-Si:H. Arrangements (a) and (b) are explained in the text. $T_s \sim T_B(S)$.

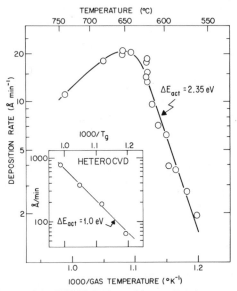

FIG. 5. Arrhenius plot of the HOMOCVD growth rate. For these experiments $T_s < 200°C$. The inset illustrates what happens when $T_s = T_g$ and the heterogeneous reaction prevails.

properties. The homogeneous decomposition of SiH_4 occurs via Eq. (1) with an activation energy of $\Delta E_{act}(gas) = 2.35$ eV mol^{-1} (Purnell and Walsh, 1966), whereas $\Delta E_{act}(surface) = 1-1.5$ eV mol^{-1}, depending on conditions (Bryant, 1979). The two regimes are clearly evident in HOMOCVD kinetic experiments shown in Fig. 5. The heterogeneous contribution to the growth process is effectively quenched for $T_s < 350°C$, and the deposition rate becomes independent of substrate temperature (Scott et al., 1981a).

HOMOCVD is practiced using gas temperatures $T_g = 650-700°C$ and $P_0(SiH_4) = 1-5$ Torr. Reactor geometries shown in Fig. 4 are the most convenient to use, as they require only minor modification of existing CVD and LPCVD systems. Figure 4a shows the original HOMOCVD arrangement, where the substrates face the gas stream (Scott et al., 1981a), whereas Fig. 4b depicts an arrangement in which the substrates are oriented in the direction opposing the gas stream (Meyerson et al., 1983). The main factors limiting the growth rate, regardless of specific configuration, are (1) SiH_4 depletion at the reactor wall; (2) homogeneous nucleation, and (3) a small temperature gradient at the substrate surface. Effects (1) and (2) are responsible for the turnover in HOMOCVD growth rate at $T_g \cong 650°C$ (Fig. 5). Depletion lowers the concentration of SiH_4 in the reactor, and therefore the SiH_2 flux (which itself is also depleted radially to the wall), whereas operation too close to the homogeneous nucleation point allows the film-forming intermediates to be taken up into higher silane molecules. As will be discussed in Part V, Section 2, factor (3) further aggravates this situation. Growing concentrations of higher silanes presage the formation of Si_xH_y clusters that comprise the homogeneously produced particles. In this regime, where the growth rate actually falls with increasing temperature (Scott *et al.*, 1981a), silicon hydride particles form in the gas and are swept toward the growing film. If the gas pressure is increased slightly, the resulting deposit will be composed entirely of a nonadhering "dust." The conditions under which this occurs depend on $P_0(SiH_4)$, T_s, T_g, and reactor geometry. In practice, we have found that the SiH_4 instability point with respect to homogeneous nucleation can be predicted approximately by the following relation:

$$\ln P_{max} = \frac{19.3 \times 10^3}{T} - 17.7,$$

where P_{max}(Torr) is the highest pressure that can be used at a given gas temperature (°K). This result is based directly on empirical homogeneous nucleation studies of $SiH_4 - H_2$ mixtures (Eversteijn, 1971) that we have found to be reasonably accurate (in the absence of H_2) using reactors of the type shown in Fig. 4a.

Most of the solid-state data reviewed in the following section were measured on a-Si:H films grown at impractically low rates (20–30 Å min^{-1}) over a very narrow range of conditions. However, as will be discussed, considerably higher rates have been achieved. The factors that contribute to low growth rate can be minimized by proper reactor design. Future work with HOMOCVD should obviously be directed toward solving these problems and exploring parameter space not yet exploited for a-Si:H growth. The range of experimental conditions possible, and the importance of factor (3), the temperature gradient, will become apparent from modeling and deposition experiments described subsequently.

IV. Properties of HOMOCVD Films

HOMOCVD permits the pyrolytic deposition of a-Si:H at substrate temperatures considerably lower than those possible with conventional isothermal CVD methods. Since ions, H atoms, and other radicals are not involved in the process, it is a "milder" form of deposition than the plasma method, for thermal energies (~0.1 eV) dominate the deposition chemistry. As a consequence, important differences between the two film types emerge that provide insight into growth environmental factors influencing film properties and make the HOMOCVD method potentially interesting technologically.

HOMOCVD film hydrogen contents are shown in Fig. 6, along with those

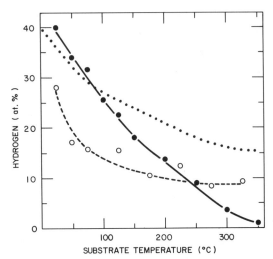

FIG. 6. Hydrogen content versus substrate temperature for HOMOCVD and plasma a-Si:H films. ●, HOMOCVD; ○, SiH₄ plasma; · · ·, Knights et al. (1979).

measured in plasma material (Scott *et al.*, 1982, 1983a). A comparison under an identical set of deposition conditions is obviously not possible, so that the data were taken from films deposited under conditions of similar growth rate ($\sim 0.5\text{Å sec}^{-1}$), corresponding to the limit of lowest rf power for plasma material. This should most realistically expose factors leading to compositional differences between the two methods. Results obtained at higher rf power are also shown in Fig. 6 (Knights, 1979).

At low substrate temperatures, films prepared in the low-power rf plasma exhibit lower C_H than HOMOCVD a-Si:H, with the point of identical composition occurring at $C_H \sim 10\%$ and $T_s \sim 250°C$. An increase in rf power displaces C_H (plasma) toward C_H (HOMOCVD), thus moving the crossing point to higher C_H and lower T_s. This could be due to a combination of greater growth rate and a larger concentration of H atoms in the higher power plasma environments (Scott *et al.*, 1983a). A significant feature is that HOMOCVD compositions are identical to the low-power plasma film values for $T_s \sim 250°C$. Further, the films are compositionally uniform despite the large temperature gradients present at the growth interface (Scott *et al.*, 1982). As a result, film properties compare favorably with plasma material deposited in the "optimum" T_s range between 200 and 300°C. Si–H bonding configurations in the network are also nearly identical for the two film types prepared in this regime (Scott *et al.*, 1981a,c, 1982).

In addition to the effects discussed in Part II, other changes are occurring in film properties as the growth temperature is lowered below those used for

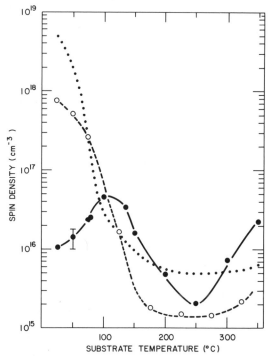

FIG. 7. Spin defect densities ("dangling" bonds) versus substrate temperature for HO-MOCVD and plasma a-Si:H films. ●, HOMOCVD; ○, SiH₄ plasma; · · ·, Fritsche *et al.* (1978).

the plasma deposition of photovoltaic material. The most significant one is shown in Fig. 7, where the EPR spin densities (N_s) are plotted for HO-MOCVD and plasma-deposited films (Scott *et al.*, 1982, 1983a; Fritzsche *et al.*, 1978). All film types display $N_s = 3 \pm 2 \times 10^{15}$ cm^{-3} in the optimum growth regime near $T_s \sim 250°$C, but as the substrate temperature drops the spin densities increase together until $T_s \sim 100°$C is reached. At this point, HOMOCVD films exhibit maximum N_s, which decreases on the low-temperature side toward $N_s = 10^{16}$ cm^{-3} as $T_s \to 25°$C. Plasma films continue to exhibit an increase in spin density as T_s drops toward room temperature, despite the fact that C_H is also increasing (Fig. 6). Clearly, dangling-bond creation mechanisms are present in plasma growth that are either muted, or do not exist, in the HOMOCVD environment at such low substrate temperatures. Dangling-bond creation has been associated with H-etching reactions of energetic ions, radicals, and hydrogen atoms in the plasma (Scott *et al.*, 1982). Although several types of defects are introduced during growth, or in the annealing period that follows, dangling bonds comprise the

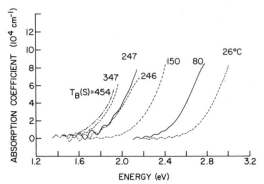

FIG. 8. Absorption coefficient versus energy for HOMOCVD a-Si:H films deposited at the substrate temperatures shown.

majority of defects contributing to the midgap density of states (Jackson and Amer, 1982). These states limit the photocurrent of actual devices by acting as traps for free carriers and serve as centers for nonradiative recombination processes in photoluminescence (Street, 1981).

As a consequence of low spin densities and optical band gaps (E_0) extending from 1.5 to 2.6 eV, HOMOCVD films exhibit unusual photo-

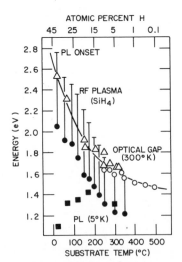

FIG. 9. Photoluminescence peak energy (●) and optical band gap (○, △) versus substrate temperature for HOMOCVD a-Si:H films on different substrates. Plasma film data are shown for comparison (■). Measurements taken at 5°K. The PL onset is given at the top of the vertical bars.

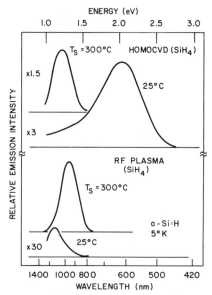

FIG. 10. Photoluminescence emission intensity versus wavelength (energy) for HOMOCVD and plasma films prepared at the temperatures shown. Data taken at 5°K.

luminescence (PL) behavior (Wolford et al., 1982, 1983a). The optical absorption data on which this gap variation is based are depicted in Fig. 8. As the gap opens (due to increasing C_H in the films), a new PL band emerges from the 1.35 eV emission line observed in plasma films deposited in the optimum range near $T_s \sim 250°C$. PL data taken at 5°K for a series of HOMOCVD films are shown in Fig. 9 (Wolford et al., 1982). The emission energies (solid circles) follow the E_0 variation with composition, reaching $E_{PL} = 2.05$ eV for room temperature-deposited films (40 at. % H). Moreover, the PL band persists, with high intensity, in the high C_H samples as they are warmed to room temperature. Typical PL results are illustrated in Fig. 10, where HOMOCVD and SiH$_4$ plasma films are compared directly. HOMOCVD material with $C_H = 40$ at. % exhibits strong, broad, featureless PL, with emission peaking in the orange-yellow region of the spectrum, whereas efficient PL is obtained at $E_{PL} \sim 1.3$ eV only from the high substrate temperature plasma-deposited sample ($T_s = 300°C$). Indeed, in the plasma films, E_{PL} is limited to 1.1–1.45 eV over a wide composition range: $10 \leq C_H \leq 30$ at. % (Fig. 9). Because of high N_s, room temperature-deposited plasma material exhibits a band greatly reduced in intensity and shifted to $E_{PL} \sim 1.2$ eV. The inverse dependence of PL intensity on N_s, previously observed in plasma films (Street et al., 1978), is actually universal

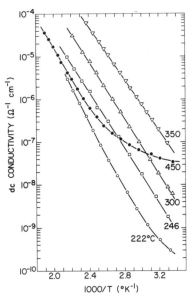

FIG. 11. Dark dc conductivity versus T^{-1} for HOMOCVD films deposited at the substrate temperatures shown.

(Wolford *et al.*, 1983a), exhibited to a much lesser degree in HOMOCVD films because of their small defect densities.

The new high-energy PL originates in the widening a-Si:H band gap with increasing C_H and has been attributed to recombination of band-tail localized excitons, possibly with strong phonon cooperation to account for its breadth (Wolford *et al.*, 1982, 1983a). The PL efficiency exceeds that of direct-gap band-to-band crystalline LED materials such as $GaAs_{0.6}P_{0.4}$ and $GaAs_{0.12}P_{0.88}$:N. Thus high-C_H HOMOCVD films are potential display materials if they can be doped. Room temperature-prepared films degrade on exposure for several months to air and humidity, but this does not occur for films deposited at $T_s \geq 50°C$ (Wolford *et al.*, 1983b). Dopability of the high-C_H material depends on the extent to which the highly cross-linked network is destroyed by the creation of Si–H bonds as the composition shifts into the region denoted SC → I in Fig. 1.

The dark dc conductivity of HOMOCVD films deposited between 220–450°C is shown in Fig. 11. Where there is deviation from linearity of ln σ_d versus T^{-1}, activation energies were extracted from the high-temperature segments of the plots (Scott *et al.*, 1982). These are shown in Fig. 12, where HOMOCVD films, like plasma-deposited material (Fritzsche, 1980), display a minimum conductivity activation energy, $\Delta E_d \sim 0.6$ eV, and maximum photoconductivity, $\sigma_p \sim 10^{-4}$ $(\Omega$ cm$)^{-1}$, near $T_s \sim 250°C$. A signifi-

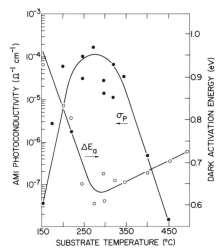

FIG. 12. AM1 photoconductivity and dark activation energy for HOMOCVD films versus growth temperature.

cant difference exists, however, because the Staebler–Wronski effect (Staebler and Wronski, 1977) is negligibly small in HOMOCVD material as shown in Fig. 13 (Scott *et al.*, 1982). HOMOCVD films deposited between $150 \leq T_s \leq 450°C$ are all similar in this respect. Comparable results have been obtained in conventional CVD films prepared from Si_2H_6 and higher silane sources (Akhtar *et al.*, 1982). The absence of Staebler–Wronski effects in CVD films could be connected with growth chemical factors that influence the nature and density of various defects such as weak and dangling bonds (Dersch *et al.*, 1981; Pankove and Berkeyheiser, 1980). On

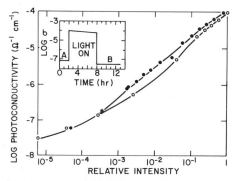

FIG. 13. Photoconductivity versus relative light intensity for a HOMOCVD film deposited at $T_s = 300°C$. The inset shows conductivity behavior versus time in the absence of light (state A, ●) and after exposure to 150 mW cm^{-2} for the period shown (state B, ○).

the other hand, the concentration and bonding state of impurities may be involved (Fritzsche, 1980). The presence of ions and radicals in the plasma environment, but not in CVD, could contribute to both possible sources of light-induced instability (Scott *et al.*, 1982).

The ability to decouple gas and film temperatures in HOMOCVD should lead to very different doping behavior. This is because gas temperature and dopant/source relative partial pressures can be adjusted to create the highest dopant precursor concentration. Although the ability to "activate" the dopant source in this way is also inherent in the plasma technique, the complexity of gas-phase reactions, combined with gas-surface chemistry involving ions, radicals, and atoms, may be deleterious to achieving a high concentration of active dopants in the film. Similarly, dopant and film source are at the same temperature in conventional CVD, and relative thermal stabilities become a paramount issue, especially at temperatures below 500°C. That there are significant differences among the various deposition methods is clearly shown in Figs. 14 and 15 (Meyerson *et al.*, 1983; Scott *et al.*, 1983b). The large shift between the solid (HOMOCVD) and dotted (plasma) curves is due to different abscissae, as only the gas-phase dopant/Si ratios were known for the latter (Spear and LeComber, 1976). Higher HOMOCVD film conductivities are evident for both *n*- and *p*-type material ($T_s = 275°C$). Disilane-deposited *n*-type CVD films ($T_s = 430°C$) exhibit maximum σ_d 20 times lower at a 10% gas phase doping level (Akhtar *et al.*, 1982). Results for *p*-type CVD films have not yet

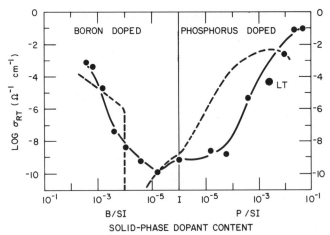

FIG. 14. Room temperature dark conductivity versus solid-state dopant–silicon ratio for HOMOCVD and plasma *p*- and *n*-type films. The plasma data are plotted against gas-phase dopant–silicon ratio. Point LT is discussed in the text. I = intrinsic film. ●, HOMOCVD; – – –, Spear (1976).

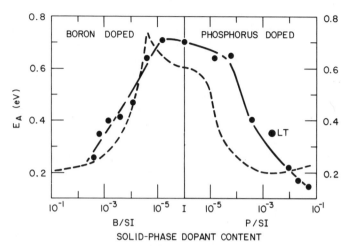

FIG. 15. Dark activation energy versus dopant–silicon ratio in the solid for HOMOCVD and in the gas phase for plasma films. ●, HOMOCVD; ---, Spear (1976).

appeared. On the other hand, doped a-Si : H prepared by CVD from PH_3-SiH_4 and $B_2H_6-SiH_4$ mixtures at temperatures between 500 and 600°C display room temperature dark conductivities ~ 0.1 $(\Omega\,cm)^{-1}$, quite comparable to HOMOCVD films (Taniguchi *et al.*, 1978; Magarino *et al.*, 1982). However, partial crystallization could be induced by high growth temperatures and the presence of boron or phosphorus. Raman spèctra prove that heavily doped HOMOCVD layers are amorphous (Meyerson *et al.*, 1983; Scott *et al.*, 1983b).

HOMOCVD films having relatively high C_H can be doped. The point denoted (LT) in Figs. 14 and 15 corresponds to a film containing 30 at. % H ($T_s = 75°C$). Dopability of high-C_H materials would allow their use as conducting window layers in photovoltaic cells, an application for which B-doped a-$(Si_{1-x}C_x)$: H is presently employed (Tawada *et al.*, 1982a). These plasma-deposited films display $E_0 = 1.8-2.2$ eV, $\Delta E_d \sim 0.55$ eV, and σ_d (25°C) $\sim 10^{-7}$ $(\Omega\,cm)^{-1}$, depending on the value of x (Tawada *et al.*, 1982b). In contrast, film (LT) exhibits $E_0 = 2.0$ eV, $\Delta E_d = 0.34$ eV, and room temperature conductivities approaching 10^{-4} $(\Omega\,cm)^{-1}$ (Meyerson *et al.*, 1983). Comparable values can be achieved on the *p*-type side (Scott *et al.*, 1983b), but dopability is lost as T_s approaches room temperature. Either the films have become too poorly cross-linked (SC → I transition, Fig. 1) or dopants will not activate into the required chemical state at low T_s. The transition from dopant activity to inactivity may occur abruptly (Scott *et al.*, 1983b).

A less explored facet of HOMOCVD is that tradeoffs in experimental parameters such as $P_0(\text{SiH}_4)$, T_g, and T_s can be used to effect changes in film composition, structure, morphology, and dopability. For example, the number of Si–Si vis-à-vis Si–H bonds might be optimized with neglible composition shift by increasing T_g while lowering T_s. As discussed in the following section, this could change the relative concentration of various gas-phase precursors. Alternatively, conditions that favor the formation of higher silanes [e.g., lower T_g and higher $P_0(\text{SiH}_4)$] could be combined with low substrate temperature to change the precursor mixture at the growth interface and thereby influence the variety of local bonding configurations. Dopant precursor activation and/or surface reaction could also be manipulated in this manner. At the present time very little is known about how such excursions in parameter space can be used to modify local order and structure so as to alter one physical property without major expense to the others. This flexibility and control is a primary advantage of HOMOCVD, compared to conventional CVD, as an electronic materials preparation technique.

V. Chemistry of a-Si:H Growth by HOMOCVD

1. Gas-Phase Reactions

The kinetic and mechanistic features of SiH_4 pyrolysis in the early stages of reaction are well established and will serve as a foundation for understanding and predicting film precursors. When SiH_4 gas is pumped into a heated reactor under conditions of the HOMOCVD experiment, the following sequence of reactions occurs:

$$\text{SiH}_4 \underset{k_{-1}}{\overset{k_1}{\rightleftharpoons}} \text{SiH}_2 + \text{H}_2, \tag{1}$$

$$\text{Si}_2\text{H}_6 \underset{k_{-2}}{\overset{k_2}{\rightleftharpoons}} \text{SiH}_2 + \text{SiH}_4, \tag{2}$$

$$\text{Si}_3\text{H}_8 \underset{k_{-3}}{\overset{k_3}{\rightleftharpoons}} \text{SiH}_2 + \text{Si}_2\text{H}_6, \tag{3}$$

$$\text{Si}_4\text{H}_{10} \underset{k_{-4}}{\overset{k_4}{\rightleftharpoons}} \text{SiH}_2 + \text{Si}_3\text{H}_8. \tag{4}$$

The presence of higher silanes in the reaction mixture is due to the insertion of the primary SiH_2 intermediate into the respective "parent" silane, that is, to the reactions having rate constants k denoted with negative subscripts. Thus in addition to molecular hydrogen, Si_2H_6 is a primary product of SiH_4 pyrolysis (Purnell and Walsh, 1966). It is produced by SiH_2 radicals from reaction (1) inserting into Si–H bonds of SiH_4 molcules, reaction (−2), making up the bulk of the gas. Disilane reacts further in (-3) to form Si_3H_8. The concentrations of higher silanes, Si_nH_{2n+2}, quickly become smaller as n increases. In this treatment the reaction sequence has been cut off at $n = 4$ because only trace amounts of Si_5H_{10} isomers are present in SiH_4 HO-MOCVD gas mixtures (Scott and Plecenik, 1981). However, since the rate constants k_{-n} are very large (John and Purnell, 1973) it is especially important to maintain experimental conditions such that higher silanes, the likely precursors to homogeneous nucleation, are not stabilized.

The silylene $[SiH_2]$ concentration in the gas would be useful to know because it provides a basis for HOMOCVD deposition rate calculations. Assuming a static reactor and neglecting diffusion to the wall, the complete steady-state expression for the SiH_2 concentration in the earliest stages of reaction can be written

$$[SiH_2] = \frac{k_1[SiH_4]^{3/2} + \sum_{n=2}^{\infty} k_n[Si_nH_{2n+2}]}{k_{-1}[H_2] + \sum_{n=1}^{\infty} k_{-(n+1)}[Si_nH_{2n+2}]}, \tag{5}$$

where concentrations are denoted by [ ] and the SiH_4 reaction order is taken from the kinetics in the unimolecular fall-off regime at higher pressure (35–230 Torr) (Purnell and Walsh, 1966). Neglecting silanes with $n > 2$, Eq. (5) reduces to (Scott *et al.,* 1981c):

$$[SiH_2] = \left(\frac{k_1}{k_{-2}}\right)[SiH_4]^{1/2} + \left(\frac{k_2}{k_{-2}}\right)\frac{[Si_2H_6]}{[SiH_4]}. \tag{6}$$

An order of magnitude estimate of $[SiH_2]$ can be made under assumed static conditions because in the later stages of reaction $[Si_2H_6]/[SiH_4]$ reaches a constant value of 0.027, independent of temperature and pressure (Purnell and Walsh, 1966). For $T_g = 650°C$ and $P_0[SiH_4] = 1.5$ Torr, a value of $(SiH_2) = 4.3 \times 10^{12}$ cm^{-3} is obtained using published Arrhenius factors and activation energies (Purnell and Walsh, 1966; Bowrey and Purnell, 1971; John and Purnell, 1973). This result accounts for nearly half of the observed a-Si : H deposition rate if plug flow, the simplest modification of the static case, is assumed for the dynamics of a HOMOCVD reactor. In this approximation we ignore the removal of SiH_2 by diffusion to the reactor wall. The calculation is nevertheless instructive, for it shows that nearly all the silylene in the steady state is derived from Si_2H_6 [the second term in (6)], despite the large insertion rate constant k_{-2}. Expressed another way, the formation of Si_2H_6 is a major channel for SiH_2 loss. If for some reason this loss could not be recovered (i.e., $k_2/k_{-2} \rightarrow 0$), only $\sim 10^{11}$ radicals cm^{-3} would be available for growth at 650°C. The success of HOMOCVD for a-Si : H deposition is a direct consequence of the fact that higher silane products of SiH_4 decomposition are themselves less kinetically and thermodynamically stable than monosilane.

Figure 16 shows a more complete calculation of the various product concentrations in static SiH_4 pyrolysis under HOMOCVD conditions: $P_0(SiH_4) = 1.5$ Torr and $T_g = 650°C$. The results were obtained by integration of the rate equations for the processes described by Eqs. (1)–(4), that is, for those up to and including tetrasilane (Scott *et al.*, 1983c). An iterative scheme was used in the computer calculations and rate constants calculated from published Arrhenius factors and activation energies (Purnell and Walsh, 1966; Bowrey and Purnell, 1971; John and Purnell 1973). The results indicate that steady state is reached in less than 1 sec (≤ 0.1 sec at 700°C), and large amounts of Si_2H_6, Si_3H_8, and Si_4H_{10} are formed in the gas. Moreover, the SiH_2 concentration, more accurately given in Fig. 16, is now 1.9×10^{13} cm^{-3}, or twice that required to explain the ~ 20 Å min^{-1} deposition rate at 650°C. Detailed reactor dynamical calculations have been carried out to assess the important loss mechanisms in the HOMOCVD geometry of Fig. 4a (Scott *et al.*, 1983b, 1983c).

Kinetic modeling for the static case (Fig. 16) bears out the predictions and interpretation given previously on the basis of Eqs. (5) and (6); namely, SiH_2 intermediates reach a high steady state concentration because less stable Si_2H_6 and higher silanes are formed and decomposed in the gas. Potentially

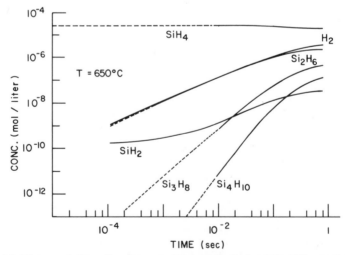

FIG. 16. Time evolution of various gas-phase products in static SiH_4 pyrolysis under HOMOCVD conditions $P_0(SiH_4) = 1.5$ Torr and $T_g = 650°C$. The steady-state SiH_2 concentration is used for a-Si:H growth rate calculations as described in the text.

important HOMOCVD gas reactions have been ignored, however, because kinetic information is lacking:

$$Si_2H_6 \rightleftarrows SiH_3SiH + H_2, \tag{7}$$

$$Si_3H_8 \rightleftarrows SiH_3SiH + SiH_4, \tag{8}$$

$$Si_3H_8 \rightleftarrows Si_2H_5SiH + H_2. \tag{9}$$

In addition, reactions involving the formation–decomposition of higher silanes with $n > 3$ are also possible. Shock-initiated Si_2H_6 pyrolysis studies between 850 and 1000°K (2300–2700 Torr) show (7) to be a major decomposition mode under those conditions (Dzarnoski *et al.*, 1982). Kinetic parameters are not available for the less important reactions (8) and (9). A calculation using the shock tube parameters is possible for reaction (7), taking the steady state (collision-stabilized) value for the Si_2H_6 concentration. The results suggest that as much as 20% of the reactive intermediates could be present as monosilyl–silylene, SiH_3–SiH, under HOMOCVD conditions at 650°C. This is likely to be a lower limit because SiH_2 and SiH_3–SiH concentrations in the gas depend on their relative rates of production from vibrationally excited disilane, and this is difficult to ascertain. The involvement of SiH_3–SiH could also be significant in conventional CVD of a-Si:H from higher silanes (Gau *et al.*, 1981a,b), possibly explaining the presence of SiH_2 stretching vibrations in the IR spectra of such films prepared at $T_s \sim 450°C$ (Akhtar *et al.*, 1982). Since this method is carried out at fairly high pressure, homogeneous decomposition of Si_2H_6 may be the rate-determining step; however, surface initiation of this dissociation reaction also cannot be discounted.

Another gas-phase reaction of possible importance is the decomposition of SiH_2 by "dimerization":

$$2SiH_2 \rightarrow SiH_3SiH^* \rightarrow SiHSiH + H_2. \tag{10}$$

The activation energy for (10) lies between 2.2–2.4 eV, and the reaction should be strongly exothermic (Newman *et al.*, 1979; Dzarnoski *et al.*, 1982) Although (10) contributes little to HOMOCVD gas-phase chemistry for $T_g < 700°C$, it could be important at all temperatures in the surface processes determining film composition, structure, and defect density.

The necessity of a sharp temperature gradient at the growth interface in HOMOCVD is evident from the preceding discussion. The process by which SiH_2 intermediates reach the substrate could be dubbed a "silane train." SiH_2 radicals jump aboard SiH_4 molecules, reaction (-2), at close to gas kinetic rates to create vibrationally excited disilane ($Si_2H_6^*$). A fraction of $Si_2H_6^*$ relaxes to the ground state, while the remaining molecules eliminate the silylene "passenger" via reaction (2). Through many repetitions of this process, SiH_2 intermediates effectively "diffuse" to the substrate, where they engage in film-forming reactions. (Those reaching the hot reactor wall are, of course, deposited there.) If the temperature gradient normal to substrate, $\Delta T/\Delta z \sim (T_g - T_s)/\Delta z$, is not steep, silylenes formed in the hot gas will be removed from the deposition zone in the guise of Si_2H_6 (and higher silane) molecules; that is, they remain passengers on the "train" and are pumped out. In terms of Eq. (6), $k_2/k_{-2} \to 0$ as $T_g \to T_s$, and the formation of cold Si_2H_6 near the substrate depletes the SiH_2 concentration.

2. SURFACE CHEMISTRY

Since the SiH_2 film precursor "diffusing" to the substrate contains 67 at. % hydrogen, the attainment of considerably lower film hydrogen content suggests a rich surface chemistry. The dependence of C_H on substrate temperature (Fig. 6), taken with IR structural data (Scott *et al.*, 1981a), led us to the model shown in Fig. 17. At the very lowest growth temperatures, SiH_2 vapor species couple on the substrate to produce a film with C_H approaching 67 at. %, that is, that of polysilylene, $(SiH_2)_n$, shown in Fig. 1. This radical recombination reaction (Fig. 17A) would proceed with essentially zero activation energy. The fact that plasma films prepared at

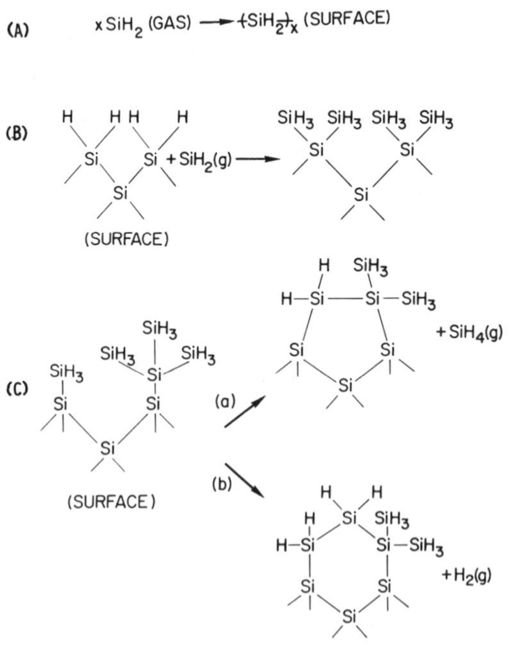

FIG. 17. Postulated sequence of surface reactions in a-Si:H growth from SiH_2 radicals.

$T_s = -125°C$ contain ~ 58 at. % hydrogen (Knights, 1979) is consistent with this expectation. In addition to polymerization, as the temperature is raised toward $25°C$, silylene can insert into surface Si–H bonds, as shown in Fig. 17B. Taken alone, this low activation energy process produces surface silyl (SiH_3) groups.

Each newly formed surface SiH_3 group contains three Si–H bonds, and any or all of these can undergo insertion by incoming SiH_2 radicals. Clearly, for steric reasons the structure cannot grow continuously in this manner. Further, C_H is diminishing with increasing substrate temperatures (Figs. 1 and 6). Hydrogen elimination and Si–Si cross-linking reactions must then be occurring to create the five-, six-, and seven-membered rings of the a-Si:H network, as depicted in Fig. 17C. A scheme based on the reaction or deactivation of surface silyl groups formed also by SiH_2 insertions has been proposed to describe how hydrogen incorporation occurs during plasma film growth (Kampas and Griffith, 1981b). Insofar as gas kinetic studies have correctly established SiH_2 as the primary intermediate in silane pyrolysis, it is reasonable to postulate the surface insertion reaction, Fig. 17B, as the initial step in a-Si:H growth at all but the lowest T_s. The cross-linking reactions that follow in Fig. 17C are more speculative. Whatever their nature, they are sufficiently rapid and complete to reduce C_H below 1 at. % for $T_s > 350°C$. Further, the thermal stability of a-Si:H grown at low T_s ($\sim 25°C$) by both HOMOCVD and plasma methods implies that the growth reactions are efficient at hydrogen removal. Films can be heated $50-100°$ hotter than their temperature of deposition before H evolution becomes measurable (Biegelsen et al., 1979). Thus it is tempting to suggest that T_s-dependent hydrogen and/or SiH_4 elimination reactions, which cross-link the network and determine C_H, proceed out of a transition state created by SiH_2 insertion; that is, Figs. 17B and 17C are concurrent. Otherwise it is necessary to consider the formation of a new intermediate on the surface following the initial insertion (Scott et al., 1983a), or postulate reactions of surface-stabilized silyl groups that have no precedent in SiH_4 gas-phase chemistry.

Recent silane plasma kinetic studies provide clues to the nature of the growth reactions (Longeway et al., 1982). On the basis of nitric oxide quenching experiments, 80–90% of the decomposed SiH_4 produces SiH_3, and the remainder SiH_2. SiH_3 radicals could abstract hydrogen from surface Si–H bonds, as well as attach to the resulting sites to continue the growth process. This scenario is appealing because it invokes a common surface intermediate (SiH_3) in both plasma and HOMOCVD despite what appear to be substantially different gas-phase chemistries (Scott et al., 1983a). A common mechanism evolving out of identical surface species creates a connection between the two techniques making it easier to understand the similarity of film properties for $T_s \geq 150°C$ (Part IV). Of course, the two deposition methods could ultimately be understood to provide very different mechanistic paths to a-Si:H.

VI. Future Prospects

There are several advantages to HOMOCVD for a-Si:H preparation. Because the method is not isothermal, relatively unreactive gases such as silane, requiring high temperatures to decompose, can be made to deposit films at low substrate temperatures. Unlike the plasma environment, there are no potentially damaging effects due to energetic ions or radiation. Further, an understanding of the pyrolysis mechanism provides clues to the chemistry of film growth. The potential usefulness of the method therefore requires a few final remarks concerning its applicability to large-scale a-Si:H thin film processing.

If HOMOCVD is to become a useful processing tool, high growth rates must be achieved while avoiding both homogeneous nucleation and SiH_4 depletion at the reactor wall. Table I presents the results of calculations and experiments designed to demonstrate that high growth rates are possible without attendant homogeneous nucleation (Scott et al., 1983c). The deposition rates at various T_g were calculated using the same iterative scheme developed for Fig. 16; that is, they are based solely on steady-state SiH_2 concentrations. The silane pressure at each T_g was taken to be no larger than its value at the homogeneous nucleation threshold, and depletion was ignored. Results of the calculations for $T_g = 650$ and $700°C$, where the wall reactions deplete less than $\sim 15\%$ of the SiH_4, are roughly twice the observed rates. The much larger discrepancy at $T_g = 800°C$ occurs because depletion

TABLE I

HOMOCVD DEPOSITION RATES[a]

Gas temp. (°C)	$P_{max}(SiH_4)$[b] (Torr)	$P_0(SiH_4)$[c] (Torr)	Flow rate (sccm)	Deposition rate (Å min⁻¹)	
				Calculated	Observed
650	25	1.5	10	45	20
700	8.5	1.5	10	200	110
800	1.3	0.9	6	900	210
900	0.3	—	10	11800	—

[a] From Scott et al. (1983c).

[b] Pressure at which homogeneous nucleation is calculated to occur (Eversteijn, 1971).

[c] Operating pressure.

is serious at this temperature ($\sim 80\%$). Nevertheless, rates of several hundred angstroms per minute are possible while operating below the homogeneous nucleation threshold; indeed, values over 200 Å min^{-1} have been measured at $T_g = 800°C$. The calculation for $T_g = 900°C$ serves only to show that considerably higher rates are possible if depletion can be arrested.

Silane loss at the reactor wall is clearly the major obstacle to more practical utilization of HOMOCVD in a-Si: H processing. Chemical engineering design principles can be applied to minimize the wall area used to heat the gas. Alternatively, the hot wall could be eliminated entirely. An example of the latter approach is the use of CO_2 laser radiation to pump SiH_4 vibrational transitions to the thermal dissociation threshold (Bilenchi *et al.*, 1982). Actually a moderate "bias" (and substrate) temperature between $280-400°C$ is applied and growth rates up to 400 Å min^{-1} have been obtained. Presumably, preheating the gas helps to ensure that k_2/k_{-2} remains large enough to avoid homogeneous nucleation and to keep the silane "train" running right up to the substrate surface.

The implementation of the various approaches such as those above for large-scale processing must await demonstration of improved HOMOCVD a-Si: H device performance. Until that time, HOMOCVD will continue to be useful in studies of material properties and their optimization under the considerably milder conditions of thermal growth chemistry.

ACKNOWLEDGMENT

The work described was partially supported by Solar Energy Research Institute under subcontract ZZ-0-9319.

REFERENCES

Akhtar, M., Dalal, V. L., Ramaprasad, K. R., Gau, S., and Cambridge, J. A. (1982). *Appl. Phys. Lett.* **41,** 1146.
Biegelsen, D. K., Street, R. A., Tsai, C. C., and Knights, J. C. (1979). *Phys. Rev. B: Condens. Matter* [3] **20,** 4839.
Bilenchi, R., Gianinoni, I., and Musci, M (1982). *J. Appl. Phys.* **53,** 6479.
Bowrey, M., and Purnell, J. H. (1971). *Proc. R. Soc. London, Ser. A* **321,** 341.
Brodsky, M. H., Frisch, M. A., Ziegler, J. F., and Lanford, W. A. (1977a). *Appl. Phys. Lett.* **30,** 561.
Brodsky, M. H., Cardona, M., and Cuomo, J. J. (1977b). *Phys. Rev. B: Condens. Matter* [3] **16,** 3556.
Bryant, W. A. (1979). *Thin Solid Films* **60,** 19.
Dersch, H., Stuke, J., and Beichler, J. (1981). *Appl. Phys. Lett.* **38,** 456.
Dzarnoski, J., Rickbarn, S. F., O'Neil, H. E., and Ring, M. A. (1982). *Organometallics* **1,** 1217.
Eversteijn, F. C. (1971). *Philips Res. Rep.* **26,** 134.
Fritzsche, H. (1980). *Sol. Energy Mater.* **3,** 447.
Fritzsche, H., Tsai, C. C., and Persans, P. (1978). *Solid State Technol.* **13,** 55.

148 BRUCE A. SCOTT

Fritzsche, H., Tanielian, M., Tsai, C. C., and Gaczi, P. J. (1979). *J. Appl. Phys.* **50**, 3366.
Gau, S. C., Weinberger, B. R., Akhtar, M., Kiss, Z., and MacDiarmid, A. G. (1981a). *AIP Conf. Proc.* **73**, 63.
Gau, S. C., Weinberger, B. R., Akhtar, M., Kiss, Z., and MacDiarmid, A. G. (1981b). *Appl. Phys. Lett.* **39**, 436.
Hamakawa, Y. (1982). "Amorphous Semiconductor Technology and Devices." North-Holland Publ., Amsterdam.
Hey, P., Raouf, N., Booth, D. C., and Seraphin, B. O. (1981). *AIP Conf. Proc.* **73**, 58.
Hirose, M., Taniguchi, M., and Osaka, Y. (1977). *Amorphous Liq. Semicond., Proc. Int. Conf., 7th, 1977* p. 352.
Hirose, M., Hamasaki, T., Mishima, Y., Kurata, H., and Osaka, Y. (1981). *AIP Conf. Proc.* **73**, 10.
Jackson, W. B., and Amer, N. M. (1982). *Phys. Rev. B* **25**, 5559.
John, P., and Purnell, J. H. (1973). *J. Chem. Soc., Faraday Trans. I* **69**, 1455.
Kampas, F. J., and Griffith, R. W. (1981a). *J. Appl. Phys.* **52**, 1285.
Kampas, F. J., and Griffith, R. W. (1981b). *Appl. Phys. Lett.* **39**, 407.
Knights, J. C. (1979). *Jpn. J. Appl. Phys.* **18**, *Suppl.* **18-1**, 101.
Knights, J. C., and Lujan, R. J. (1979). *Appl. Phys. Lett.* **35**, 244.
Knights, J. C., Lucovsky, G., and Nemanich, R. J. (1978). *Philos. Mag. [Part] B* **37**, 467.
Kshirsagar, S. T., and Lannin, J. S. (1982). *Phys. Rev. B* **25**, 2916.
Longeway, P. A., Estes, R. D., and Weakliem, H. A. (1982). *J. Phys. Chem.* (to be published).
Magarino, J., Kaplan, D., Friederich, A., and Deneuville, A. (1982). *Philos. Mag. [Part] B* **45**, 285.
Meyerson, B. S., Scott, B. A., and Wolford, D. J. (1983). *J. Appl. Phys.* **54**, 1461.
Milleville, H., Fuhs, W., Demond, F. J., Mannsperger, H., Müller, G., and Kalbitzer, S. (1979). *Appl. Phys. Lett.* **34**, 173.
Nakashita, T., Taniguchi, M., and Osaka, Y. (1981). *Jpn. J. Appl. Phys.* **20**, 471.
Newman, C. G., Ring, M. A., and O'Neal, H. E. (1979). *J. Am. Chem. Soc.* **100**, 5945.
Oguz, S., and Paesler, M. A. (1980). *Phys. Rev. B: Condens. Matter* [3] **22**, 6213.
Pankove, J. I., and Berkeyheiser, J. E. (1980). *Appl. Phys. Lett.* **37**, 705.
Perrin, J., and Delafosse, E. (1980). *J. Phys. D* **13**, 759.
Purnell, J. H., and Walsh, W. (1966). *Proc. R. Soc. London, Ser. A* **293**, 543.
Ring, M. A. (1977). *In* "Homoatomic Rings, Chains and Macromolecules of Main-Group Elements" (A. L. Rheingold, ed.), pp. 121–163. Elsevier, Amsterdam.
Scott, B. A., and Plecenik, R. M. (1981). Unpublished results.
Scott, B. A., Brodsky, M. H., Green, D. C., Kirby, P. B., Plecenik, R. M., and Simonyi, E. E. (1980). *Appl. Phys. Lett.* **37**, 725.
Scott, B. A., Plecenik, R. M., and Simonyi, E. E. (1981a). *Appl. Phys. Lett.* **39**, 73.
Scott, B. A., Brodsky, M. H., Green, D. C., Plecenik, R. M., Simonyi, E. E., and Serino, R. (1981b). *AIP Conf. Proc.* **73**, 6.
Scott, B. A., Plecenik, R. M., and Simonyi, E. E. (1981c). *J. Phys. (Orsay, Fr.)* **42**, C4-635.
Scott, B. A., Reimer, J. A., Plecenik, R. M., Simonyi, E. E., and Reuter, W. (1982). *Appl. Phys. Lett.* **40**, 973.
Scott, B. A., Reimer, J. A., and Longeway, P. A. (1983a). *J. Appl. Phys.* (to be published).
Scott, B. A., Olbricht, W. L., Reimer, J. A., Meyerson, B. S., and Wolford, D. J. (1983b). *J. Non-Cryst. Solids* (to be published).
Scott, B. A., Olbricht, W. L., Meyerson, B. S., Reimer, J. A., and Wolford, D. J. (1983c). *J. Vac. Sci. Tech. A* (to be published).
Spear, W. E., and LeComber, P. G. (1976). *Philos. Mag. [Part] B* **33**, 935.
Staebler, D. L., and Wronski, C. R. (1977). *Appl. Phys. Lett.* **31**, 292.

Street, R. A. (1981). *Phys. Rev. B* **23**, 861.

Street, R. A., Knights, J. C., and Biegelsen, D. K. (1978). *Phys. Rev. B* **18**, 1880.

Tachibana, K., Tadokaro, H., Hauma, Y., and Urano, Y. (1982). *J. Phys. D* **15**, 177.

Taniguchi, M., Hirose, M., and Osaka, Y. (1978). *J. Cryst. Growth* **45**, 126.

Taniguchi, M., Hirose, M., Hamasaki, T., and Osaka, T. (1980). *Appl. Phys. Lett.* **37**, 787.

Tawada, Y., Kondo, M., Okamoto, H., and Hamakawa, Y. (1982a). *Sol. Energy Mater.* **6**, 299.

Tawada, Y., Tsuge, K., Kondo, M., Okamoto, H., and Hamakawa, Y. (1982b). *J. Appl. Phys.* **53**, 5273.

Turban, G., Catherine, Y., and Grolleau, B. (1982). *Plasma Chem. Plasma Process.* **2**, 61.

Walsh, R. (1981). *Acc. Chem. Res.* **14**, 246.

Wolford, D. J., Scott, B. A., Reimer, J. A., and Bradley, J. A. (1982). *Physica* **117E, 118B**, 920.

Wolford, D. J., Reimer, J. A., and Scott, B. A. (1983a). *Appl. Phys. Lett.* **42**, 369.

Wolford, D. J., Scott, B. A., and Reimer, J. A. (1983b). Unpublished results.

Yonezawa, F. (1981). "Fundamental Physics of Amorphous Semiconductors," Springer Solid State Sci. Ser. 25. Springer-Verlag, Berlin and New York.

Characteristics of Silane Plasma

SEMICONDUCTORS AND SEMIMETALS, VOL. 21A

CHAPTER 8

Chemical Reactions in Plasma Deposition

Frank J. Kampas

DIVISION OF METALLURGY AND MATERIALS SCIENCE
BROOKHAVEN NATIONAL LABORATORY
UPTON, NEW YORK

I. Introduction and Overview

Glow-discharge deposition is the most complicated technique used for the production of hydrogenated amorphous silicon (a-Si : H). However, it is also the method that has produced the most efficient solar cells. The connection between deposition parameters and the properties of the film produced is indirect, as shown in Table I. Electrons in the discharge are accelerated by the electric field and undergo both elastic and inelastic collisions with the atoms and molecules present. The electron energy distribution is therefore determined by the types of atoms and molecules present in the discharge, their concentration, and the magnitude and frequency of the electric field. Inelastic collisions with atoms and molecules can ionize them, replacing electrons lost to the discharge. Inelastic collisions with molecules can raise them to unstable excited states that dissociate. In silane discharges SiH_n ($n = 0-3$) free radicals and atomic hydrogen are produced as well as ions. Increasing the electric field increases both the number and average energy of the electrons.

The SiH_n species produced polymerize on the substrate to produce the

153

TABLE I

THE EFFECT OF DEPOSITION PARAMETERS ON THE
FILM GROWTH PROCESS[a]

Deposition parameters	Film growth process

Electrode geometry
Voltage
Frequency
Gas composition
Flow rate
Pressure
 	
Substrate temperature
	Plasma Composition:
	  Stable molecules
	  Reactive fragments
	  Electrons
	  Ions
	  Photons
	↓
	Flux of particles onto film
	↓
	Composition, structure, and morphology
	of film
	↓
	Optoelectronic properties of film

[a] Arrows in the diagram represent causal relationships.

a-Si:H film. Rapid hydrogen elimination occurs from activated species produced by the reaction of these SiH_n species with the film surface. Slower, thermally activated hydrogen elimination can also occur if the substrate temperature is high enough. Atomic hydrogen can break and hydrogenate Si–Si bonds as well as remove surface hydrogen from the film. Ion bombardment also removes hydrogen from the film, probably by providing activation energy for hydrogen elimination.

II. Historical Background

One of the first efforts to understand the chemical reactions occurring in silane discharges was a study of the kinetics of silane decomposition in a dc discharge (Nolet, 1975). The first-order rate constant was found to depend approximately linearly on current and to decrease with increasing total pressure for a 10% silane in argon mixture. These results were interpreted in terms of the effects of pressure and current on the electron concentration and temperature, in a model in which the initial step involved an electron–molecule collision.

Brodsky (1977) pointed out that the films deposited from silane glow discharges contain hydrogen, and he discussed the effects of pressure on the type of film deposited. At low pressure, surface reactions predominate, whereas gas-phase polymerization is more important at higher pressures.

Turban *et al.* (1979) published a study of the kinetics of a-Si:H deposition in a silane – helium rf discharge. They used a model of the deposition process in which electron – molecule collisions produce neutral free radicals. These are transported to the walls where they undergo surface polymerization. The electron concentration was measured by microwave interferometry as a function of power, and the deposition rate was measured as a function of power and position in the deposition system. The experimental results were explained by the model, and the silane reactivity was found to be an order of magnitude higher than that of methane.

In a paper presented at the Eighth International Conference on Amorphous and Liquid Semiconductors (Cambridge, MA, 1979), Knights (1980) discussed the nucleation and growth of a-Si : H films and proposed that his observation could be explained if the films grow from SiH_2 and SiH_3. These species would be expected to have higher surface mobility than more unsaturated species such as atomic Si.

At the same conference, Griffith et al. (1980) reported that the optical emission spectra of silane rf glow discharges show the presence of the species Si, SiH, H, H_2, and some impurities. There was also a paper by Kocian (1980), who measured the electron concentration and energy distribution in a dc silane discharge at various conditions and discussed the changes in the properties of the deposited films.

Subsequently, several papers appeared reporting on the emission spectra measured during the plasma deposition of a-Si : H under a variety of conditions. Kampas and Griffith (1980) observed emission from SiO, N_2, and SiCl when oxygen, nitrogen, or chlorosilanes were present in the discharge and discussed the effects of these impurities on film properties. Perrin and Delafosse (1980) examined the emission from SiH in silane – hydrogen rf discharges and concluded that the rotational and vibrational temperatures of the emitting excited state of SiH are abnormally high.

Taniguchi et al. (1980) examined the effect of a magnetic field on the emission spectra and films deposited from rf silane discharges. Emission from SiH, H, and H_2 decreased with increasing magnetic field due to a decrease in the electron temperature. Infrared absorption studies of the deposited films showed increased monohydride concentration and decreased dihydride concentration with the application of the magnetic field. Hamasaki et al. (1980) reported that crystalline phosphorus-doped silicon could be deposited from a silane – hydrogen rf discharge at substrate temperatures below 200°C at low flow rates in the presence of a magnetic field. Under those conditions the emission from SiH is very weak compared to emission from H and H_2, and emission from doubly excited H_2 states is absent. This was ascribed to reactive hydrogen attacking the film surface and increasing the surface mobility of silicon atoms.

Mass spectrometry was another technique used to study the chemical reactions occurring during the plasma deposition of a-Si : H. Drevillon *et al.* (1980) examined the deposition and mass spectra from a low-pressure (<5 mTorr) dc multipole silane discharge. The predominant ion was found to be SiH_3^+ and the contribution of ions to the deposition was found to decrease rapidly with pressure, implying that for discharges around 100 mTorr pressure, neutral species in the plasma are responsible for the deposition. Haller (1980) examined the ions formed in rf silane discharges in the pressure range of 17–100 mTorr and found ions containing as many as seven silicon atoms whose concentration decreased by a constant ratio with an increasing number of silicon atoms. Turban *et al.* (1980) also observed ions with as many as seven silicon atoms in rf discharges in silane–hydrogen and silane–helium mixtures. Although they were not able to observe neutral-free radicals directly, they concluded that such radicals must be responsible for film deposition, since the ion flux is one to two orders of magnitude too low to explain the deposition rate. The mass spectrum of a H_2 discharge showed that H can etch a-Si : H, and studies of silane–helium–deuterium discharges implied that atomic H or D in the plasma can be incorporated in the film. This was believed to be most important at high power or low flow rate.

The discovery by Scott *et al.* (1980) that the glow-discharge deposition of a-Si : H proceeds at a considerably higher rate from disilane than from monosilane led to their proposal that the film grows from SiH_2. They argued that glow-discharge processes are analogous to the thermal decomposition of the various silanes, which are known to produce SiH_2 and to proceed at a higher rate for disilane than monosilane. In order to account for the fact that a-Si : H films prepared by glow discharge have a smaller hydrogen content than SiH_2, they proposed that a rapid surface reaction eliminates silane and molecular hydrogen.

At the Tetrahedrally Bonded Amorphous Semiconductor Conference (Carefree, Arizona, 1981) Kampas and Griffith (1981a,b) presented a model of the glow-discharge deposition of a-Si : H in which SiH_2 inserts into a Si–H bond on the growing film surface, producing an activated group that can eliminate H_2 to produce a divalent silicon. This divalent silicon can then insert into a nearby Si–H bond to cross-link the structure. At the same conference Hirose *et al.* (1981) reported on the optical emission spectra and deposited film properties from silane–hydrogen rf glow discharges as a function of flow rate. They found that the Si and SiH emission intensities increased with flow rate, as did the deposition rate and the monohydride concentration in the films. However, the dihydride concentration decreased. These results were interpreted in terms of the reactions

$$SiH + H \rightarrow SiH_2,$$

$$SiH + H \rightarrow Si + H_2$$

occurring on the film surface. Dalal *et al.* (1981) presented a model of the deposition of a-Si:(H,F) from SiF_4 and H_2 that explained many of the properties of that material in terms of reactive-ion etching occurring during deposition. Reimer and Knights (1981) presented data obtained by nuclear magnetic resonance on the concentration and distribution of hydrogen in a-Si:H films prepared under a variety of deposition conditions. They found a positive correlation between deposition rate and hydrogen content that they explained in terms of increased strain in the lattice producing more hydrogenated structures. They also found that cathode films have a lower concentration of polysilane regions. This was explained in terms of "scouring" of the film surface by silicon-containing ions (also see Reimer *et al.,* 1981).

A paper by Kampas and Griffith (1981c) on the emission spectrum of the silane discharge showed that the emitting excited states of Si and SiH are produced directly by the electron impact decomposition of the silane molecule, explaining the high rotational and vibrational temperatures observed for SiH by Perrin and Delafosse (1980).

Stable molecule and ion concentrations in SiH_4–SiD_4–He and SiH_4–H_2–D_2 rf glow discharges were measured by Turban *et al.* (1981, 1982) using mass spectrometry. The data were explained using known reactions between SiH_2, SiH_3, and H occurring in the discharge. A picture of the deposition mechanism was proposed in which chemisorbed SiH_2 and SiH_3 are incorporated into the growing film with some H_2 elimination, and H can either be incorporated in the film or etch it. A paper by the same group (Catherine *et al.,* 1981) on the reactions in the deposition of a-SiC:H from SiH_4–CH_4 mixtures concluded that the two gases break down independently to make SiH_2, SiH_3, CH_2, and CH_3, which react to grow the film and make disilane and various hydrocarbons.

At the Ninth International Conference on Amorphous and Liquid Semiconductors (Grenoble, France, July 2–8, 1981) Matsuda *et al.* (1981) presented a simplified model of the glow-discharge deposition of a-Si:H and a-Si:(H,F) based on the strengths of the bonds between the atoms in the reactive species that occur in the plasma. Using results from a cross-field deposition system, Hotta *et al.* (1981) reported on the effect of the substrate potential on the optical emission spectrum of the discharge and the properties of deposited films. The emission spectrum only from the plasma near the substrate was affected by the bias voltage. The dark conductivity and photoconductivity, hydrogen bonding structure, and doping efficiency of the films were all found to depend on substrate potential. Veprek *et al.* (1981) discussed the deposition of amorphous and microcrystalline silicon films from silane–hydrogen discharges in terms of the departure of the system from chemical equilibrium, the forward reaction being deposition and the reverse reaction etching of the film by atomic hydrogen. The film is amorphous when the system is far from equilibrium and microcrystalline when the deposition is near partial chemical equilibrium (also see Wagner and Vepřek, 1982).

Optical spectroscopy continued to be a useful technique for studying the glow-discharge deposition of a-Si:H. Knights *et al.* (1982) reported on the spectrum of a silane discharge in the infrared. Emission and absorption were observed for both SiH_4 and SiH molecules. Vibrational and rotational temperatures of 2000 and 485°K were deduced for SiH and 850 and 300°K for SiH_4. The species SiH_2 and SiH_3 were not observed, probably because the lower symmetry of those molecules splits the lines into more components, decreasing their intensities. In a separate publication, Perrin and Schmitt (1982) reported the cross sections for production of excited Si, Si^+, SiH, SiH^+, and H by electron impact excitation of silane in the energy range 17–68 eV. Tachibana *et al.* (1982) measured the time dependence of the optical absorption of atomic Si in a pulsed rf silane–argon discharge in order to obtain the Si concentration and diffusion constant. They concluded that 70% of the deposition in their system is due to atomic Si.

Drawing an analogy with the gas-phase reaction of atomic H with Si_2H_6, Kampas (1982) proposed that H can remove a bonded H to produce a dangling bond, break a Si–Si bond to produce a dangling bond, or passivate a dangling bond during the deposition of a-Si:H by glow discharge or reactive sputtering. This explains qualitatively the change in the substrate temperature dependence of the film hydrogen content when H_2 is present in the discharge.

A study of the neutral and ion chemistry in ethylene–silane discharges by Catherine *et al.* (1982) showed that the major neutral species produced by electron impact dissociation of C_2H_4 and C_2H_3 and C_2H_2. This result and the results obtained by the same group on silane–methane discharges (Catherine *et al.*, 1981) are in agreement with the results of a study of the infrared spectra of a-SiC:H films produced by glow discharge, which concluded that material produced from silane–methane mixtures contains CH_3 groups, whereas material produced from silane–ethylene mixtures contains C_2H_5 groups (Tawada *et al.*, 1982).

Recently, Perrin *et al.* (1982) have measured the electron impact dissociation cross sections of silane and disilane from their ionization energies to 100 eV. The most probable decomposition mechanism of disilane involves formation of monosilane and other fragments.

At the Solar Energy Research Institute Amorphous Materials Subcontractors Review Meeting (Alexandria, Va., 1982), the results of two mass spectrometric studies of silane discharges were presented. Gallagher detected neutral free radicals and found the SiH_3 was the most common product (Robertson *et al.*, 1983). Longeway measured the effect of NO, a

SiH_3 scavenger, on the production of higher silanes by the discharge (Longeway *et al.*, 1983). He concluded that SiH_3 is produced at a rate nine times higher than that of SiH_2 and noted that there was no film deposition under conditions of complete SiH_3 scavenging.

III. Primary Processes in the Discharge

An inelastic collision between an electron and a molecule can raise the molecule to a higher rotational, vibrational, or electronic state. Excited electronic states of molecules are often unstable because an electron has been removed from a bonding orbital and transferred to an antibonding orbital. The excited molecule may decompose, ionize, or both:

$$e^- + X \rightarrow A + B + e^-, \tag{I}$$

$$e^- + X \rightarrow X^+ + 2e^-, \tag{II}$$

$$e^- + X \rightarrow A^+ + B + 2e^-. \tag{III}$$

The rates of these reactions are given by Bell (1974):

$$R_i = n_e k_i[X], \tag{1}$$

where n_e is the electron concentration, k_i is the rate constant for the reaction, and $[X]$ is the concentration of X. The rate constant is given by

$$k_i = (2/m_e)^{1/2} \int E^{1/2} f(E) \sigma_i(E) \, dE, \tag{2}$$

where m_e is the electron mass, $f(E)$ is the normalized electron energy distribution, and $\sigma_i(E)$ is the cross section for the reaction as a function of energy. The electron energy distribution is not necessarily Maxwellian, although it is often assumed to be for simplicity. Typical values for n_e are $1 \times 10^9 - 1 \times 10^{12}$ cm^{-3}; the average electron energy is usually several electron volts.

1. DISSOCIATION

Enthalpies of formation of the various SiH_n ($n = 0-3$) fragments are known, so it is possible to calculate the energy required to produce them (Turban, 1981):

$$e^- + SiH_4 \rightarrow SiH_2 + H_2 + e^- \quad (\Delta H = 2.2 \text{ eV}), \tag{IV}$$

$$SiH_3 + H + e^- \quad (\Delta H = 4.0 \text{ eV}), \tag{V}$$

$$Si + 2H_2 + e^- \quad (\Delta H = 4.2 \text{ eV}), \tag{VI}$$

$$SiH + H_2 + H + e^- \quad (\Delta H = 5.7 \text{ eV}). \tag{VII}$$

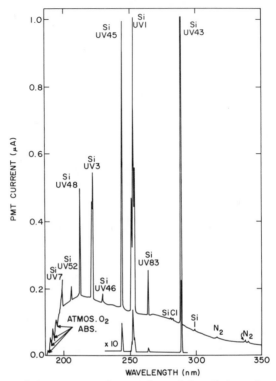

FIG. 1. Optical emission spectrum of a pure silane, rf glow discharge from 180 to 350 nm. The emission from SiCl at 280 nm is due to the impurity SiH_3Cl in the silane. The atomic Si lines are labeled using the notation of Moore (1967).

Another important reaction is the dissociation of H_2, which is a product of the deposition process and is sometimes used to dilute the silane:

$$e^- + H_2 \rightarrow 2H + e^- \qquad (\Delta H = 4.5 \text{ eV}). \qquad \text{(VIII)}$$

Furthermore, there are reactions similar to reactions (IV), (VI), and (VII) in which 2H is produced instead of H_2. These require 4.5 eV more energy for each H_2 dissociated.

As discussed earlier, the rates of the primary reactions depend on the electron concentration and energy distribution. Some information about how these factors depend on deposition conditions can be obtained using optical emission spectroscopy, since the emitting excited states of some of the species observed are formed by dissociation of silane (Kampas and Griffith, 1981c, Perrin and Schmitt, 1982) as shown below:

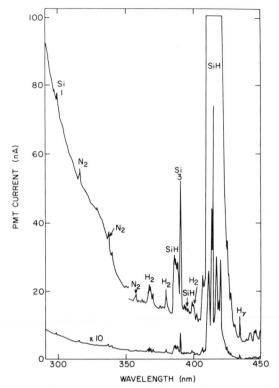

FIG. 2. Optical emission spectrum of a pure silane discharge from 300 to 450 nm.

$$e^- + SiH_4 \rightarrow Si^* + 2H_2 + e^- \qquad (\Delta H = 9.5 \text{ eV}), \qquad \text{(IX)}$$

$$e^- + SiH_4 \rightarrow SiH^* + H_2 + H + e^- \qquad (\Delta H = 8.9 \text{ eV}). \qquad \text{(X)}$$

An emission spectrum of a silane discharge is shown in Figs. 1–3.

The fact that the Si* and the SiH* observed in the silane glow discharge are produced primarily by reactions (IX) and (X) rather than by excitation of ground state Si and SiH was deduced from a study of the rf power dependences of the emission intensities from Si, SiH, H, H_2, and N_2 in a silane discharge containing a small fixed concentration of N_2 (Kampas and Griffith, 1981c). Log–log plots of the emission intensities from Si, SiH, H, and H_2 versus the emission intensity from N_2 gave straight lines with slopes of 0.84, 0.78, 1.92, and 1.85, respectively (see Fig. 4). Since the formation of N_2^* is a one-electron process,

$$e^- + N_2 \rightarrow N_2^* + e^- \qquad (\Delta H = 11.1 \text{ eV}), \qquad \text{(XI)}$$

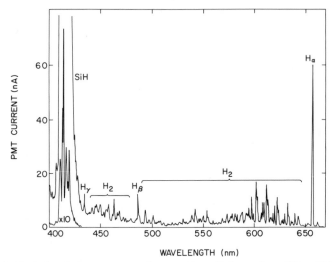

FIG. 3. Optical emission spectrum of a pure silane discharge from 400 to 650 nm.

formation of Si* and SiH* had to be one-electron processes also to account for the slopes of 0.84 and 0.78. The slopes were found to be larger for species with larger excitation energy because the electron energy distribution shifts to higher energy with increasing power.

Since the slopes corresponding to the emission intensities from H_2^* and H* were near 2, it was concluded that both of these species arise from excitation from the ground state. However, Perrin and Schmitt (1982) have concluded that H* can be formed from SiH_4 or H_2:

$$e^- + SiH_4 \rightarrow H^* + SiH_3 + e^-, \tag{XII}$$

$$e^- + H_2 \rightarrow H^* + H + e^-. \tag{XIII}$$

However, emission from H_2^* is a result of excitation of ground state H_2. Emission from H* and SiH* is not a measure of the concentration of H and SiH but rather a measure of the concentration of H_2 and SiH_4.

Although the optical emission does not give information about the ground state concentrations of species other than H_2, it is useful for studying how electron concentration and energy distribution change with deposition conditions, as illustrated earlier. Emission spectroscopy has shown that the average electron energy increases with rf power. Emission spectroscopy has also demonstrated that electron concentration decreases and average electron energy increases with increasing silane concentration in argon–silane discharges at fixed power and total pressure. This explains why the deposition rate of a-Si:H from argon–silane mixtures is not proportional to the silane fraction (Kampas, 1983).

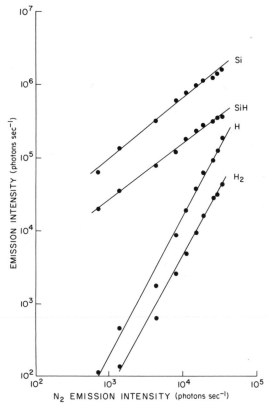

FIG. 4. Log–log plot of emission intensity of Si, SiH, H, and H_2 versus N_2 emission intensity as the rf power level is varied in a silane discharge with a small amount of added N_2.

2. IONIZATION

The appearance potentials of the positive ions formed from the ionization of silane and disilane [reactions (XIV)–(XVII)] have been measured by Potzinger and Lampe (1969) and the ionization cross sections have been measured approximately by Turban et al. (1982). The SiH_4^+ ion is unstable and, therefore, not observed (Gordon, 1978).

$$e^- + SiH_4 \rightarrow SiH_2^+ + H_2 + 2e^- \qquad (\Delta H = 11.9 \text{ eV}), \qquad \text{(XIV)}$$

$$SiH_3^+ + H + 2e^- \qquad (\Delta H = 12.3 \text{ eV}), \qquad \text{(XV)}$$

$$Si^+ + 2H_2 + 2e^- \qquad (\Delta H = 13.6 \text{ eV}), \qquad \text{(XVI)}$$

$$SiH^+ + H_2 + H + 2e^- \qquad (\Delta H = 15.3 \text{ eV}). \qquad \text{(XVII)}$$

Turban et al. (1982) have calculated the rate constants for these reactions as a function of average electron energy, assuming a Maxwellian electron energy distribution.

IV. Secondary Reactions

From the previous discussion it is clear that the species formed by the electron impact dissociation of silane are SiH_n ($n = 0-3$), H, H_2, and SiH_n^+ ($n = 0-3$). Of these SiH_2 and SiH_3 are probably formed in the largest quantities. With the exception of H_2, all of the species are very reactive. Their gas phase chemistry will be discussed in some detail in order to understand what reactions they might undergo in the plasma, and as a guide to the reactions involved in film growth.

3. NEUTRAL REACTIONS

The absorption spectrum of the SiH_2 molecule has been observed in a pulsed silane–hydrogen discharge and shows that the molecule is bent and has a singlet ground state (Dubois, 1968). Its chemistry is fairly well established because it can be produced by the pyrolysis of monosilane (Newman et al., 1979) and by the pyrolysis of disilane (Bowrey and Purnell, 1971):

$$SiH_4 \rightarrow SiH_2 + H_2 \qquad (\Delta H = 2.2 \text{ eV}), \qquad (XVIII)$$

$$Si_2H_6 \rightarrow SiH_4 + SiH_2 \qquad (\Delta H = 2.1 \text{ eV}). \qquad (XIX)$$

The pyrolysis of trisilane produced SiH_3SiH as well as SiH_2 (Vanderwielen et al., 1975).

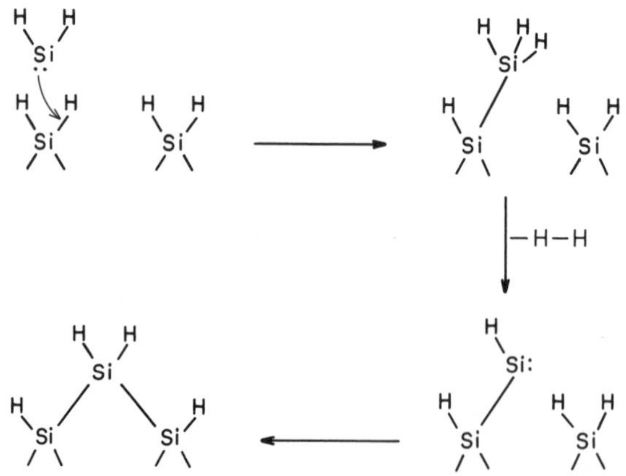

FIG. 5. Proposed mechanism of SiH_2 reaction with growing a-Si:H film and subsequent hydrogen elimination and cross-linking.

The reverse of reactions (XVIII) and (XIX) is also known:

$$SiH_2 + H_2 \rightarrow SiH_4 \qquad (\Delta H = -2.2 \text{ eV}), \qquad (XX)$$

$$SiH_2 + SiH_4 \rightarrow Si_2H_6 \qquad (\Delta H = -2.1 \text{ eV}). \qquad (XXI)$$

(John and Purnell, 1973). Reaction (XXI) is an example of a well-characterized insertion reaction into a Si–H bond as shown in Fig. 5.

The SiH_3 radical has a doublet ground state and is pyramidal in shape, according to electron spin resonance (Morehouse et al., 1966). It is rapidly formed by the reaction of atomic hydrogen with silane (Austin and Lampe, 1977):

$$H + SiH_4 \rightarrow H_2 + SiH_3 \qquad (\Delta H = -0.5 \text{ eV}) \qquad (XXII)$$

There is a similar reaction between atomic hydrogen and disilane as well as a reaction in which atomic hydrogen splits the Si–Si bond in disilane (Pollock et al., 1973):

$$H + Si_2H_6 \rightarrow H_2 + Si_2H_5, \qquad (XXIII)$$

$$H + Si_2H_6 \rightarrow SiH_3 + SiH_4 \qquad (\Delta H = -0.5 \text{ eV}). \qquad (XXIV)$$

The SiH_3 radical can react with itself in two ways: disproportionation or the formation of an activated disilane molecule (Reiman et al., 1977).

$$2 \, SiH_3 \rightarrow SiH_2 + SiH_4 \qquad (\Delta H = -1.4 \text{ eV}), \qquad (XXV)$$

$$2 \, SiH_3 \rightarrow Si_2H_6^{**}. \qquad (XXVI)$$

The notation $Si_2H_6^{**}$ indicates a disilane molecule with an internal energy of 3.2 eV (Perkins et al., 1979). This molecule breaks up [reaction (XXVII)] unless deactivated by a collision [reaction (XVIII)] (Reiman et al., 1977):

$$Si_2H_6^{**} \rightarrow SiH_2 + SiH_4, \qquad (XXVII)$$

$$M + Si_2H_6^{**} \rightarrow M + Si_2H_6. \qquad (XXVIII)$$

Reaction (XXV) can be distinguished from the sequence reaction (XXVI) and reaction (XXVII) by isotopic labeling.

It is also worth noting that SiH_3 can react with atomic hydrogen (or deuterium) (Mihelcic et al., 1974):

$$D + SiH_3 \rightarrow SiH_3D \qquad (\Delta H = -4.0 \text{ eV}). \qquad (XXIX)$$

The reactions of atomic Si and SiH are less well understood. Atomic Si is known to react with a number of molecules such as H_2, CH_4, and $SiCl_4$ (Husain and Norris, 1978). The reaction of Si with H_2 goes as follows:

$$Si + H_2 \rightarrow SiH + H. \qquad (XXX)$$

A mass spectrometric study of the photolysis of silane by 8.4-eV photons provides a model for the reactions that might occur in glow-discharge deposition from silane (Perkins *et al.*, 1979). The results were interpreted in terms of the formation of SiH_2, SiH_3, and H in the primary photolytic step:

$$hv + SiH_4 \rightarrow SiH_2 + 2H, \qquad \text{(XXXI)}$$

$$hv + SiH_4 \rightarrow SiH_3 + H. \qquad \text{(XXXII)}$$

However, it was noted that production of Si and SiH is possible energetically. The analysis of the data gave quantum yields of 0.83 and 0.17 for the production of SiH_2 and SiH_3, respectively.

An important result was that disilane and trisilane were formed simultaneously. Addition of NO, which supresses any pathways involving species with unpaired spin such as SiH_3, reduced the disilane yield to 0.53 of its initial value and reduced the trisilane yield to 0.07 of its initial value. In order to explain the prompt formation of trisilane in the presence of NO, the following reaction sequence was proposed (Perkins *et al.*, 1979):

$$SiH_2 + SiH_4 \rightarrow Si_2H_6^*, \qquad \text{(XXXIII)}$$

$$Si_2H_6^* \rightarrow SiH_3SiH + H_2, \qquad \text{(XXXIV)}$$

$$SiH_3SiH + SiH_4 \rightarrow Si_3H_8. \qquad \text{(XXXV)}$$

The species $Si_2H_6^*$ is a disilane molecule with an internal energy of 2.1 eV. Reaction (XXIV) was estimated to be slightly exothermic. The small trisilane yield from this mechanism (7% of the total) was ascribed to competition between the reaction (XXXIV) and deactivation of $Si_2H_6^*$:

$$M + Si_2H_6^* \rightarrow M + Si_2H_6. \qquad \text{(XXXVI)}$$

Gaspar (1981) has criticized reaction (XXXIV) on the grounds that trisilane formation could also be explained by the generation of Si or SiH in the primary photolysis step. Perkins *et al.* also proposed the reaction

$$Si_2H_6^{**} \rightarrow H_2 + SiH_3SiH, \qquad \text{(XXXVII)}$$

where $Si_2H_6^{**}$ is formed from two SiH_3 molecules as described earlier [reaction (XXVI)]. They concluded that most of the $Si_2H_6^{**}$ molecules decomposed according to reaction (XXXVII) under the conditions of the measurement (silane–helium mixture, 2 Torr silane partial pressure, 42 Torr total pressure).

In their mass spectrometric studies of silane discharges, Turban *et al.* (1981, 1982) found Si_2H_6 in SiH_4–He discharges and SiH_2D_2, $SiHD_3$, and all partially deuterated disilanes in SiH_4–SiD_4–He discharges. The formation of the partially deuterated silanes was ascribed to the reaction (XXIX) and to a reaction analogous to (XXV):

$$SiH_3 + SiD_3 \rightarrow SiH_mD_{4-m} + SiH_nD_{2-n} \qquad (\Delta H = 1.6 \text{ eV}). \quad \text{(XXXVIII)}$$

Disilane formation in the discharge occurred through the reaction (XXI), since the pressure was not high enough for stabilization of $Si_2H_6^{**}$ formed from SiH_3 as in reaction (XXXVI).

4. ION–MOLECULE REACTIONS

Ion–molecule reactions are faster than radical–molecule reactions. This explains why ions containing as many as seven silicon atoms are found in silane discharges (Haller, 1980; Turban $et\ al.$, 1980), whereas the largest stable neutral molecule found in the silane discharge is trisilane. Henis $et\ al.$ (1972) have studied the reactions of SiH_n^+ ($n = 0-3$) with SiH_4:

$$Si^+ + SiH_4 \rightarrow Si_2H_2^+ + H_2, \qquad \text{(XXXIX)}$$

$$SiH^+ + SiH_4 \rightarrow Si_2H_3^+ + H_2, \qquad \text{(XL)}$$

$$SiH^+ + SiH_4 \rightarrow Si_2H^+ + 2H_2, \qquad \text{(XLI)}$$

$$SiH_2^+ + SiH_4 \rightarrow Si_2H_2^+ + 2H_2, \qquad \text{(XLII)}$$

$$SiH_2^+ + SiH_4 \rightarrow Si_2H_4^+ + H_2, \qquad \text{(XLIII)}$$

$$SiH_2^+ + SiH_4 \rightarrow SiH_3 + SiH_3^+, \qquad \text{(XLIV)}$$

$$SiH_3^+ + SiH_4 \rightarrow Si_2H_5^+ + H_2, \qquad \text{(XLV)}$$

$$SiH_3^+ + SiH_4 \rightarrow Si_2H_3^+ + 2H_2, \qquad \text{(XLVI)}$$

$$SiH_3^+ + SiH_4 \rightarrow SiH_4 + SiH_3^+. \qquad \text{(XLVII)}$$

The H^- transfer reaction (XLVII) was established by using isotopically labeled ions.

Reaction (XLIV) is the reason that SiH_3^+ is the ion with the highest concentration in the silane discharge (Drevillon $et\ al.$, 1980; Haller, 1980; Turban $et\ al.$, 1980) even though SiH_2^+ is the main product of the ionization of silane (Turban $et\ al.$, 1982). The higher silane ions are the result of sequential reactions (Haller, 1980; Turban $et\ al.$, 1980, 1982).

V. Film Growth Reactions

Reactive species in the plasma include both radicals: SiH_n ($n = 0-3$), atomic H, and ions: $Si_xH_y^+$, and H^+. The ion flux onto the surface of the growing film is too small to account for the rate of film growth (Turban $et\ al.$, 1980), which strongly suggests that the silicon-containing species responsible for film growth are the neutral silane fragments SiH_n. Recent results have indicated that SiH_3 is the neutral free radical produced in the largest quantity in silane discharges (Robertson $et\ al.$, 1983) and that the a-Si:H film can be deposited from it (Longeway $et\ al.$, 1983).

The reactions that might be expected to occur between SiH_2 or SiH_3 and the growing film surface can be described, based on the gas-phase chemistry of these species. The thermal decomposition of silane and disilane produces SiH_2 only via reactions (XVIII) and (XIX). Therefore, the film growth reactions for a-Si : H produced by chemical vapor deposition (CVD) are also applicable to the reactions of SiH_2 in glow-discharge deposition.

The first step in the reaction of a chemisorbed SiH_2 with the growing amorphous film surface should be the addition of the SiH_2 across a Si–H bond on the film surface, in analogy to reaction (XXXIII).

$$\text{Si-H} + SiH_2 \longrightarrow \text{Si-SiH}_3^*. \tag{XLVIII}$$

The asterisk indicates that the surface complex has approximately 2 eV internal energy.

The processes that occur after the initial insertion have been the subject of some speculation. It is clear that there is rapid elimination of H_2 because films grown by CVD and glow discharge have much smaller hydrogen content than the SiH_2 molecule. The author has proposed that H_2 be eliminated immediately (Kampas and Griffith, 1981a,b) in analogy to reaction (XXXIV), which was proposed by Perkins *et al.* (1979) to explain the results of their mass spectrometric study of silane photolysis:

$$\text{Si-SiH}_3^* \longrightarrow \text{Si-SiH} + H_2. \tag{XLIX}$$

The reaction competes with deactivation of the excited complex:

$$\text{Si-SiH}_3^* \longrightarrow \text{Si-SiH}_3. \tag{L}$$

The divalent silicon atom formed as a result of the H_2 elimination in reaction (XLIX) next attacks an adjacent Si–H bond, in this model:

$$\text{Si-SiH} + HSi \longrightarrow \text{Si-SiH}_2\text{-Si}. \tag{LI}$$

The entire sequence of reactions is shown in Fig. 5.

Scott has proposed a similar scheme in which H_2 or SiH_4 can be eliminated in fast surface reactions (Scott *et al.*, 1981) as shown in Fig. 6. These concerted reactions involve atoms attached to two different silicon atoms. Heating polysilane $(SiH_2)_n$ in a vacuum is known to produce various silanes as well as hydrogen (Schwarz and Heinrich, 1935). Furthermore, there is some evidence that hydrogen evolution from already deposited films involves hydrogens on different silicon atoms rather than two hydrogens on the same silicon atom (John *et al.*, 1980).

As far as their overall effect on the film hydrogen content is concerned, the two models cannot be easily distinguished. Therefore, the mechanism of hydrogen elimination following SiH_2 insertion must be regarded as an open question.

The species SiH_3 has an unpaired spin and should react with unpaired spins on the growing film surface (dangling bonds)

$$\underset{\displaystyle \geq}{}Si\cdot + SiH_3 \longrightarrow \underset{\displaystyle \geq}{}Si-SiH_3^{**}, \tag{LII}$$

where $\geq Si-SiH_3^{**}$ has an internal energy of about 3 eV.

Reaction (LII) has been written in analogy to reaction (XXVI). Hydrogen elimination and deactivation reactions for $\geq Si-SiH_3^{**}$ can be written that are similar to the reactions of $\geq Si-SiH_3^{*}$ [(XLIX) and (L)]. Another possible reaction of SiH_3 with the a-Si:H film surface is hydrogen abstraction, resulting in a dangling bond:

$$\underset{\displaystyle \geq}{}Si-H + SiH_3 \longrightarrow \underset{\displaystyle \geq}{}Si\cdot + SiH_4. \tag{LIII}$$

A film can grow from SiH_3 alone, according to the reactions (LII) and (LIII). Perkins *et al.* (1979) have concluded that the deposition on the walls of their photolysis chamber is due to species with unpaired spins such as SiH_3 and not divalent silicon species that are too reactive to reach the walls. Brodsky and Haller (1980) and Knights (1981) have deposited a-Si:H films by reacting atomic H with SiH_4, which produces SiH_3 [reaction (XXII)]. Longeway has shown that NO, which scavenges SiH_3, prevents film deposition in a dc discharge at a silane pressure of 0.5 Torr (Longeway *et al.*, 1983).

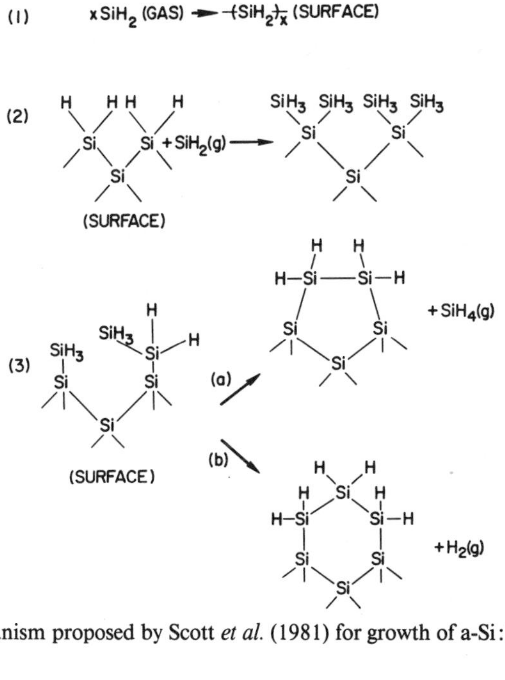

FIG. 6. Mechanism proposed by Scott *et al.* (1981) for growth of a-Si:H from SiH_2.

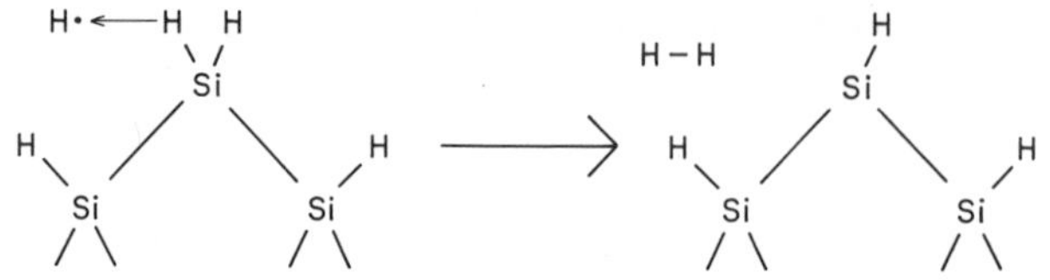

FIG. 7. Reaction of atomic H with a-Si:H, removing a H to produce a dangling bond.

That is probably too high a silane pressure for any SiH_2 to reach the substrate.

The reactions of Si or SiH with the growing film surface are more a matter of speculation than in the case of SiH_2 and SiH_3, whose gas-phase chemistry is better understood. However, it seems likely that activated groups result on the surface that may lead to hydrogen elimination. Films of a-Si:H deposited by reactive sputtering grow from Si and H (Moustakas et al., 1981; Tiedje et al., 1981). However, there has not been a complete study of the substrate temperature dependence of the film hydrogen content using this deposition method. Therefore, it is difficult to draw any conclusions about the possible mechanisms of hydrogen elimination. As far as glow-discharge deposition is concerned, Si and SiH probably react before reaching the substrate under most deposition conditions.

There is considerable evidence that atomic hydrogen in the plasma reacts with the growing film surface (see historical background section). The reactions of atomic hydrogen with silane and disilane [reactions (XXIII), (XXIV), and (XXIX)] can serve as a guide to the reactions at the film surface (Kampas, 1982). Atomic hydrogen is expected to remove hydrogen, leaving a dangling bond (see Fig. 7); break a Si–Si bond, leaving a dangling bond (Fig. 8); or passivate a dangling bond (Fig. 9). It is interesting to note that H can both create and passivate dangling bonds. Under conditions that produce a large flux of atomic hydrogen on the film surface in comparison to the flux of silicon containing species, etching of the film will become important. This occurs when all the bonds between a surface silicon atom and the silicons beneath it are broken. Weaker bonds will be more reactive toward H so that the Si atoms most likely to be etched away are those in positions that result in strained bonds. Therefore, the fact that microcrystal-

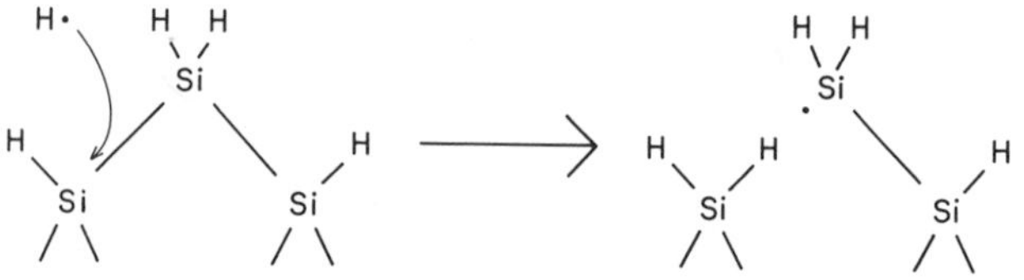

FIG. 8. Reaction of atomic H with a-Si:H, breaking a Si–Si bond.

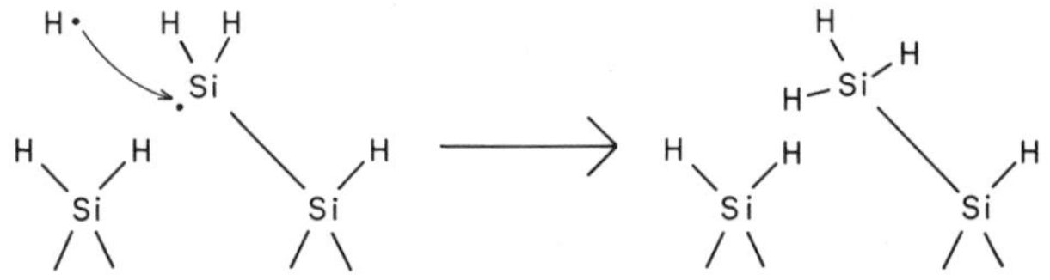

FIG. 9. Passivation of a dangling bond by atomic H.

line silicon, rather than amorphous silicon, is deposited at conditions of large silane dilution in hydrogen can be understood as being caused by etching away of surface atoms not fitting into the crystal structure.

Another factor that is known to affect the composition of the deposited a-Si:H film is ion bombardment. It is common practice in rf diode glow-discharge deposition to connect the rf "hot lead" to one electrode and connect the other electrode and the chamber to ground. This asymmetry results in the ungrounded electrode becoming negative with respect to ground (Chapman, 1980). The negative electrode is thus termed the cathode. Films grown on the cathode experience more ion bombardment than films grown on the grounded electrode (anode). Cathode films generally have less hydrogen than anode films; in addition, a smaller fraction of the hydrogen is in the form of dihydride groups or polysilane. Reimer and Knights (1981) have postulated that ion bombardment scours the growing film of polysilane regions (also see Reimer *et al.,* 1981).

VI. Reactions Involved in Doping

The fact that it is possible to dope a-Si:H *n*- or *p*-type is essential to the usefulness of this material. However, the mechanisms involved in doping are mysterious. At one time it was thought that doping was impossible in an amorphous material, since the flexibility of the amorphous network would allow dopant atoms to have their normal coordination number rather than the electrically active coordination. Actually, the dopant atoms seem to have a variety of coordination numbers so that some fraction of them are active.

Phosophine (PH_3) is normally used as an *n*-type dopant, although arsine (AsH_3) has been used as well. Extended x-ray absorption fine structure (EXAFS) studies have shown that approximately 20% of the arsenic atoms are four-fold coordinated with silicon at doping concentrations of around 1% (Hayes *et al.,* 1977).

About the only guide to what the chemistry of *n*-type doping might be is a study of the 147-nm photolysis of phosphine–silane mixtures (Blazejowski and Lampe, 1981). The primary products of PH_3 photolysis are PH, PH*,

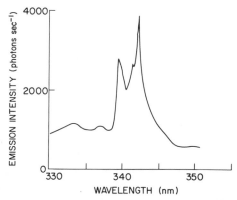

FIG. 10. Emission from PH ($A^3\pi \rightarrow X^3\Sigma$ transition) in a 1% PH_3 in SiH_4 rf discharge.

and PH_2^*. This is in agreement with the emission spectrum (see Figs. 10 and 11) of an rf phosphine–silane mixture glow discharge that showed emission from PH and PH_2 (Kampas and Griffith, 1981a). The only gaseous product of the photolysis containing both phosphorus and silicon was PH_2SiH_3. Its formation was ascribed to two reactions:

$$PH_2 + SiH_3 + M \rightarrow PH_2SiH_3 + M, \tag{LIV}$$

$$SiH_2 + PH_3 + M \rightarrow PH_2SiH_3 + M. \tag{LV}$$

where M represents any third body.

These reactions indicate what might happen on the surface of the growing film. However, not much as been done in this area at the present time.

Less is known about diborane than phosphine. The photolysis of diborane

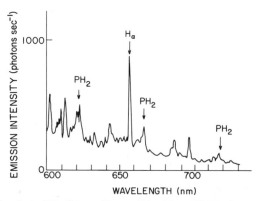

FIG. 11. Emission from PH_2 ($^2A_1 \rightarrow ^2B_1$ transition) in a 1% PH_3 in SiH_4 discharge. Peaks not labeled are due to H_2.

by photons of 185-nm light produces both B_2H_5 and BH_3 with a quantum yield of near 1 for B_2H_5 and around 0.1 for BH_3 (Kreve and Marcus, 1962). One might therefore expect that boron dimers are found in boron-doped a-Si:H. However, this remains controversial.

VII. Deposition from Gases Other Than Silane

Hydrogenated amorphous silicon has been deposited from disilane as well as silane. This has the advantage of a higher deposition rate (Scott *et al.*, 1980). However, the reason for the higher rate has not been well established. Scott *et al.* have suggested that the glow-discharge decomposition is similar to thermal decomposition, as the thermal decomposition is faster for disilane than for monosilane. The increase in glow-discharge deposition rate varies, depending on the deposition system and conditions. Increases in deposition rate by factors of 20 (Scott *et al.*, 1980) and 5 (Delahoy *et al.*, 1982) have been reported. Perrin *et al.* (1982) have measured the electron impact dissociation cross sections for silane and disilane from their ionization energies to 100 eV. The maximum values, which occur at 60 eV, are

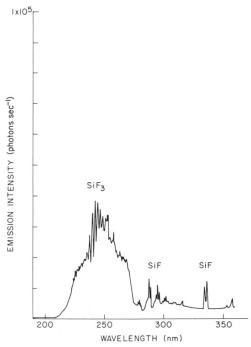

FIG. 12. Emission spectrum of $SiF_4 + H_2$ discharge from 200 to 350 nm.

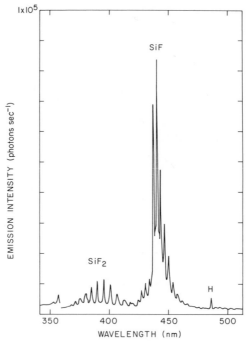

FIG. 13. Emission spectrum of a $SiF_4 + H_2$ discharge from 350 to 500 nm. The break in the curve at 360 nm is due to insertion of an order filter.

1.2×10^{-15} cm^2 and 2.6×10^{-15} cm^2 for silane and disilane, respectively. Perrin *et al.* also estimate the thresholds for decomposition of silane and disilane to be 8 eV and 7 eV, respectively. The larger cross section and lower threshold for disilane may be sufficient to explain the higher deposition rate.

The alloy a-Si:(H,F) has been deposited from $SiF_4 + H_2$ mixtures (Ovshinsky and Madan, 1978). Emission spectra from SiF_4 discharges show the presence of SiF_n ($n = 1-3$), but there is little or no deposition from SiF_4 alone (see Figs. 12 and 13). Discharges containing SiF_4 and H_2 also show emission from H and H_2. The species SiF_2 can be produced from Si and SiF_4 at high temperature. However, pyrolysis of SiF_2 produces a film with a high spin density (Janai *et al.*, 1981). It seems that deposition of a film with a low defect density requires hydrogen.

VIII. Future Research

The most fundamental question about glow-discharge deposition of a-Si:H concerns the identity of the species that result in film growth.

Arguments have been made for Si, SiH, SiH_2, SiH_3, and various combinations of those. Mass spectrometry is just now beginning to give some answers to that question. One must also wonder if there are fundamental differences between material deposited from different silicon – hydrogen free radicals. The nature of defect formation processes is also a question with important technological ramifications. Finally, the mechanism of doping is almost completely mysterious. Presumably, the techniques applied to the question of the silicon-containing free precursors will be applied to the dopant precursors as well.

ACKNOWLEDGMENT

The writing of this chapter was partially supported by the U.S. Department of Energy, Office of Basic Energy Sciences under Contract No. DE-AC02-76CH00016.

REFERENCES

Austin, E. R., and Lampe, F. W. (1977). *J. Phys. Chem.* **81**, 1134–1138.
Bell, A. T. (1974). *In* "Techniques and Applications of Plasma Chemistry" (J. R. Hollahan and A. T. Bell, eds.), p. 31. Wiley, New York.
Blazejowski, J., and Lampe, F. H. (1981). *J. Photochem.* **16**, 105–120.
Bowrey, M., and Purnell, J. H. (1971). *Proc. R. Soc. London, Ser. A* **321**, 341–359.
Brodsky, M. H. (1977). *Thin Solid Films* **40**, L23–L25.
Brodsky, M. H., and Haller, I. (1980). *IBM Tech. Disclosure Bull.* **22**, 3391–3392.
Catherine, Y., Turban, G., and Grolleau, B. (1981). *Thin Solid Films* **76**, 23–33.
Catherine, Y., Turban, G., and Grolleau, B. (1982). *Plasma Chem. Plasma Process.* **2**, 81–93.
Chapman, B. (1980). "Glow Discharge Processes," p. 156. Wiley, New York.
Dalal, V. K., Fortmann, C. M., and Eser, E. (1981). *In* "Tetrahedrally Bonded Amorphous Semiconductors" (R. A. Street, D. K. Biegelsen, and J. C. Knights, eds.), pp. 15–19. Am. Inst. Phys., New York.
Delahoy, A. E., Kampas, F. J., Corderman, R. R., Vanier, P. E., and Griffith, R. W. (1982). *Conf. Rec. IEEE Photovoltaic Spec. Conf.* **16**, 1117–1123.
Drevillon, G., Huc, J., Lloret, A., Perrin, J., de Rosny, G., and Schmitt, J. P. M. (1980). *Appl. Phys. Lett.* **37**, 646–648.
Dubois, I. (1968). *Can. J. Phys.* **46**, 2485–2490.
Gaspar, P. P. (1981). *In* "Reactive Intermediates" (M. Jones, Jr. and R. A. Moss, eds.), Vol. 2, p. 340. Wiley, New York.
Gordon, M. S. (1978). *Chem. Phys. Lett.* **59**, 410–413.
Griffith, R. W., Kampas, F. J., Vanier, P. E., and Hirsch, M. D. (1980). *J. Non-Cryst. Solids* **35–36**, 391–396.
Haller, I. (1980). *Appl. Phys. Lett.* **37**, 282–284.
Hamasaki, T., Kurata, H., Hirose, M., and Osaka, Y. (1980). *Appl. Phys. Lett.* **37**, 1084–1086.
Hayes, T. M., Knights, J. C., and Mikkelsen, J. C., Jr. (1977). *Amorphous Liq. Semicond., Proc. Int. Conf., 7th, 1977* pp. 73–77.
Henis, J. M. S., Stewart, G. W., Tripodi, M. K., and Gaspar, P. P. (1972). *J. Chem. Phys.* **57**, 389–398.

176 FRANK J. KAMPAS

Hirose, M., Hamasaki, T., Mishima, Y., Kurata, H., and Osaka, Y. (1981). *In* "Tetrahedally Bonded Amorphous Semiconductors" (R. A. Street, D. K. Biegelsen, and J. C. Knights, eds.), pp. 10–14. Am. Inst. Phys., New York.

Hotta, S., Tawada, Y., Okamoto, H., and Hamakawa, Y. (1981). *J. Phys. (Orsay, Fr.)* **42**, Suppl. **10**, C4-631–C4-634.

Husain, D., and Norris, P. E. (1978). *J. Chem. Soc., Faraday Trans. 2* **74**, 1483–1503.

Janai, M., Weil, R., Levin, K. H., Pratt, B., Kalish, R., Braunstein, G., Teicher, M., and Wolf, M. (1981). *J. Appl. Phys.* **52**, 3622–3624.

John, P., and Purnell, J. H. (1973). *J. Chem. Soc., Faraday Trans. 1* **69**, 1455–1461.

John, P., Odeh, I. M., Thomas, M. J. K., Tricker, M. J., Riddoch, F., and Wilson, J. I. B. (1980). *Philos. Mag. [Part] B* **42**, 671–681.

Kampas, F. J. (1982). *J. Appl. Phys.* **53**, 6408–6412.

Kampas, F. J. (1983). *J. Appl. Phys.* **54**, 2276–2280.

Kampas, F. J., and Griffith, R. W. (1980). *Sol. Cells* **2**, 385–400.

Kampas, F. J., and Griffith, R. W. (1981a). *In* "Tetrahedrally Bonded Amorphous Semiconductors" (R. A. Street, D. K. Biegelsen, and J. C. Knights, eds.), pp. 1–5. Am. Inst. Phys., New York.

Kampas, F. J., and Griffith, R. W. (1981b). *Appl. Phys. Lett.* **39**, 407–409.

Kampas, F. J., and Griffith, R. W. (1981c). *J. Appl. Phys.* **52**, 1285–1288.

Knights, J. C. (1980). *J. Non-Cryst. Solids* **35–36**, 159–170.

Knights, J. C. (1981). *Bull Am. Phys. Soc.* **26**, 389.

Knights, J. C., Schmitt, J. P. M., Perrin, J., and Guelachvili, G. (1982). *J. Chem. Phys.* **76**, 3414–3421.

Kocian, P. (1980). *J. Non-Cryst. Solids* **35–36**, 195–200.

Kreve, W. C., and Marcus, R. A. (1962). *J. Chem. Phys.* **37**, 419–427.

Longeway, P. A., Wiekliem, H. A., and Estes, R. D. (1983). *J. Phys. Chem.* (to be published).

Matsuda, A., Matsumura, M., Nakagawa, K., Yamasaki, S., and Tanaka, J. (1981). *J. Phys. (Orsay, Fr.)* **42**, Suppl. **10**, C4-687–C4-690.

Mihelcic, D., Potzinger, P., and Schindler, R. N. (1974). *Ber. Bunsenges. Phys. Chem.* **78**, 82–89.

Moore, C. E. (1967). "Selected Tables of Atomic Spectra, Si I," NSRDS-NBS 3, Sect. 2. Natl. Bur. Stand., Washington, D.C.

Morehouse, R. L., Christiansen, J. J., and Gordy, W. (1966). *J. Chem. Phys.* **45**, 1751–1758.

Moustakas, T. D., Tiedje, T., and Lanford, W. A. (1981). *In* "Tetrahedrally Bonded Amorphous Semiconductors" (R. A. Street, D. K. Biegelsen, and J. C. Knights, eds.), pp. 20–24. Am. Inst. Phys., New York.

Newman, C. G., O'Neil, H. E., Ring, M. A., Leska, F., and Shipley, N. (1979). *Int. J. Chem. Kinet.* **11**, 1167–1182.

Nolet, G. (1975). *J. Electrochem. Soc.* **122**, 1030–1034.

Ovshinsky, S. R., and Madan, A. (1978). *Nature (London)* **276**, 482–484.

Perkins, G. G. A., Austin, E. R., and Lampe, F. W. (1979). *J. Am. Chem. Soc.* **101**, 1109–1115.

Perrin, J., and Delafosse, E. (1980). *J. Phys. D* **13**, 759–765.

Perrin, J., and Schmitt, J. P. M. (1982). *Chem. Phys.* **67**, 167–176.

Perrin, J., Schmitt, J. P. M., de Rosny, G., Drevillon, B., Huc, J., and Lloret, A. (1982). *Chem. Phys.* **73**, 383–394.

Pollock, T. L., Sandhu, H. S., Jodhan, A., and Strausz, O. P. (1973). *J. Am. Chem. Soc.* **95**, 1017–1024.

Potzinger, P., and Lampe, F. W. (1969). *J. Phys. Chem.* **73**, 3912–3917.

Reiman, B., Matten, A., Laupert, R., and Potzinger, P. (1977). *Ber. Bunsenges. Phys. Chem.* **81**, 500–504.

Reimer, J. A., and Knights, J. C. (1981). *In* "Tetrahedrally Bonded Amorphous Semiconductors" (R. A. Street, D. K. Biegelsen, and J. C. Knights, eds.), pp. 78–82. Am. Ins. Phys., New York.

Reimer, J. A., Vaughan, R. W., and Knights, J. C. (1981). *Phys. Rev. B* **24**, 3360–3370.

Robertson, R., Hills, D., Chatham, H., and Gallagher, A. (1983). *Appl. Phys. Lett.* **43**, 544–546.

Schwarz, R., and Heinrich, F. (1935). *Z. Anorg. Chem.* **221**, 277–286.

Scott, B. A., Brodsky, M. H., Green, D. C., Kirby, P. B., Plecenik, R. M., and Simonyi E. E. (1980). *Appl. Phys. Lett.* **37**, 725–727.

Scott, B. A., Plecenik, R. M., and Simonyi, E. E. (1981). *Appl. Phys. Lett.* **39**, 73–75.

Tachibana, K., Tadokoro, H., Harima, H., and Urano, Y. (1982). *J. Phys. D* **15**, 177–184.

Taniguchi, M., Hirose, M., Hamasaki, T., and Osaka, Y. (1980). *Appl. Phys. Lett.* **37**, 787–788.

Tawada, Y., Tsuge, K., Kondo, M., Okamoto, H., and Hamakawa, Y. (1982). *J. Appl. Phys.* **53**, 5273–5281.

Tiedje, T., Moustakas, T. D., and Cebulka, J. M. (1981). *Phys. Rev. B: Condens. Matter* [3] **23**, 5634–5637.

Turban, G. (1981). Thèse d'État, Université de Nantes.

Turban, G., Catherine, Y., and Grolleau, B. (1979). *Thin Solid Films* **60**, 147–155.

Turban, G., Catherine, Y., and Grolleau, B. (1980). *Thin Solid Films* **67**, 309–320.

Turban, G., Catherine, Y., and Grolleau, B. (1981). *Thin Solid Films* **77**, 287–300.

Turban, G., Catherine, Y., and Grolleau, B. (1982). *Plasma Chem. Plasma Process.* **2**, 61–80.

Vanderwielen, A. J., Ring, M. A., and O'Neal, H. E. (1975). *J. Am. Chem. Soc.* **97**, 993–998.

Vepřek, S., Iqbal, Z., Oswald, H. R., Sarott, F. A., and Wagner, J. J. (1981). *J. Phys. (Orsay, Fr.)* **42**, *Suppl.* **10**, C4-251–C4-255.

Wagner, J. J., and Vepřek, S. (1982). *Plasma Chem. Plasma Process.* **2**, 95–107.

CHAPTER 9

Plasma Kinetics

Paul A. Longeway

RCA LABORATORIES
PRINCETON, NEW JERSEY

I. Introduction

The plasma discharge decomposition of molecules represents a uniquely challenging and complex chemical system, both mechanistically and kinetically. The investigator of such a system must not only have knowledge of the basic chemistry of the species under examination but also have a general understanding of plasma physics. In sorting out the decomposition processes, he must consider ionization, electronic excitation, and perhaps even vibrational excitation. The likelihood of these processes occurring must then be juxtaposed against a non-Boltzmann electron energy distribution, the fact that the electrons undergoing collision are then reaccelerated by an electric field that is, by the way, nonuniform throughout the plasma, and the pertinent cross sections for the processes envisioned. The electron impact processes will include electron attachment, dissociative attachment to generate negative ions and neutrals, positive ionization that additionally acts to multiply the electron density of the plasma, fragmentation to neutrals, electron–ion recombination, wall recombination, radiative recombination, and elastic and inelastic collisions leading to internal excitation of the molecule. At this point the chemical kineticist may well be tempted to throw his hands up in despair and turn to a much simpler problem. Nonetheless, considerable headway can be made against these odds.

In the following sections we will discuss the plasma decomposition

179

processes of silane (SiH_4). We will present what we consider to be a reasonable overview of the major primary dissociative mechanisms and the further reactions expected for the resulting primary fragments. These will be balanced against the known kinetic parameters and will lead, we hope, to some general conclusions about the nature of the intermediates involved in the production of amorphous Si:H materials during plasma decomposition. In addition, we will discuss briefly the implications of these conclusions for a quite different problem, namely, the surface chemistry of the film formation process. We refer the reader to the chapter in this volume by Kampas (Chapter 8) for alternate points of view and further discussion.

II. Silane Primary Decomposition Processes

We have listed as part 1 of Table I what we consider to be the major electron-impact dissociative processes operative in the silane plasma. These choices are based on the following considerations.

TABLE I

PLASMA DECOMPOSITION REACTIONS

	Reaction	ΔH (kcal mol^{-1})
1.	Primary Processes	
	a. $e^- + SiH_4 \rightarrow SiH_3^+ + H + 2e^-$	284
	b. $\rightarrow SiH_2^+ + H_2 + 2e^-$	263
	c. $\rightarrow SiH_2 + 2H + e^-$	155
	d. $\rightarrow SiH + H_2 + H + e$	131
	e. $\rightarrow Si + 2H_2 + e^-$	97
	f. $\rightarrow SiH_3 + H + e^-$	90
	g. $\rightarrow SiH_2 + H_2 + e^-$	56
2.	Secondary reactions	
	1. $SiH_3^+ + SiH_4 + M \rightarrow Si_2H_7^+ + M$	—
	2. $SiH_3^+ + SiH_4 \rightarrow Si_2H_5^+ + H_2$	-2.9
	3. $SiH_2^+ + SiH_4 \rightarrow SiH_3^+ + SiH_3$	-1
	4. $SiH_2 + SiH_4 \rightarrow Si_2H_6$	-49
	5. $SiH_3 + SiH_3 \rightarrow Si_2H_6$	-74
	6. $H + SiH_4 \rightarrow SiH_3H_2$	-17
	7. $SiH_3 + SiH_3 \rightarrow SiH_2 + SiH_4$	-36
	8. $SiH_3 + SiH_3 \rightarrow SiH_3SiH + H_2$	—
	9. $SiH_2 + SiH_4 \rightarrow SiH_3SiH + H_2$	—
	10. $SiH_3SiH: \rightarrow \cdot SiH_2SiH_2SiH_3\cdot$	$+11$
	11. $SiH_3SiH + SiH_4 \rightarrow Si_3H_8$	-62
	12. $SiH + SiH_4 \rightarrow SiH_3SiH_2$	-45
	13. $Si + SiH_4 \rightarrow SiH_3SiH$	-34

Mass spectrometric analysis of silane discharges in the "ionizer-off" mode allows direct observation of ions produced in the discharge. While ionic clusters containing from 2–5 Si atoms are observed (Weakliem, 1980), only the monosilicon ions can be considered to be primary ionic fragments, that is, fragments created directly by the electron-impact dissociation of SiH_4. The entire gamut of monosilicon ions from Si^+ to SiH_4^+ is observed in the plasma; however, there is general agreement that the most abundant of these by at least an order of magnitude are SiH_2^+ and SiH_3^+ (Weakliem, 1980; Matsuda and Tanaka, 1982; Drevillon et al., 1981). It is for this reason that we have included reactions (a) and (b) in Table I. Additionally, these reactions are observed during the γ-ray radiolysis of silane (Lampe and Schmitt, 1969).

The diradical SiH_2, produced via reactions (c) and (g) of Table I, has been directly observed using mass spectrometric techniques by both Bourquard et al. (1981) and Gallagher and Scott (1982). Bourquard et al. (1981) concluded that the flux of SiH_2 onto the walls of a discharge region was greater than that of any other radical. On the other hand, Gallagher and Scott (1983) concluded that SiH_3 had the largest flux density, followed by SiH_2. The photon–molecule reaction analogous to reaction (c) of Table I has been shown to occur to a significant extent in the vacuum UV photolysis of silane (Perkins et al., 1979) and the photon–molecule analog of reaction (g) is the known decomposition mechanism for both the pyrolysis of silane (Neudorfl et al., 1980) and the IR-laser multiphoton decomposition of silane (Lampe and Longeway, 1981). In the latter case, SiH_2 emission has not been so observed via optical emission spectroscopy (OES) (Lampe and O'Keefe, 1983). The SiH_2 emission has not been so observed in the plasma discharge, partially because the emission bands are weak and fall within the much more intense Balmer series for H_2 emission. Further discussion of OES measurements in the silane plasma can be found in the chapter in this volume by Weakliem (Chapter 10).

Reaction (d) of Table I, which generates SiH, has been included based on the observation that the strongest OES line in the silane discharge spectrum is that of SiH* (where the symbol * indicates an excited state) (Matsuda and Tanaka, 1982). Furthermore, Griffith and Kampas (1981) have demonstrated that the SiH* emission so observed is generated by a single electron-impact process.

Likewise, reaction (e) of Table I is included based on OES observation of the Si* emission lines in the discharge spectrum (Matsuda and Tanaka, 1982).

The primary generation of SiH_3, a monoradical, is included as reaction (f) of Table I based on the studies of Longeway et al. (1983) and Catalano (1983), as well as the previously mentioned study by Gallagher and Scott

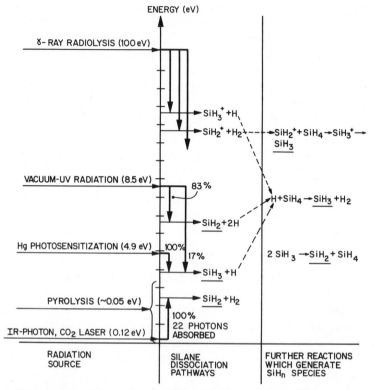

FIG. 1. Observed dissociation pathways for several radiation decompositions of monosilane. Where percentages appear, this indicates the extent of that particular decomposition pathway that is observed for a specific radiation decomposition method. The further reactions listed in the right-hand column are not scaled to the energy axis.

(1983). In the first two instances, the investigators were able to suppress film formation completely from a silane discharge and reduce the formation of disilane, Si_2H_6 [see reaction (5), part 2, Table I], by the introduction of nitric oxide (NO) to the reaction chamber. Nitric oxide is a known free-radical scavenger for SiH_3 and is inert toward singlet-state SiH_2 [Perkins *et al.,* 1979). Additionally, the appropriate analogue of reaction (f) is the known decomposition process for silane via both the H-atom-induced decomposition (Austin and Lampe, 1977) and the Hg-photosensitized decomposition of silane (Kamaratos and Lampe, 1970).

Note that all the reactions listed in part 1 of Table I also give rise to H_2 and/or H. Thus any reaction mechanism must also include the reactions of H with other species to be complete.

We have included as Fig. 1 a diagram of the reported decomposition processes for silane via several radiation sources, all of which have been referred to herein. Note specifically that the energy of the radiation source very much determines the decomposition pathway observed.

We have compiled from Table I a list of primary fragments whose further reactions with each other and silane will lead to the observed stable products of the silane discharge. These primary fragments are SiH_2^+, SiH_3^+, Si, SiH, SiH_2, SiH_3, and H. We will now examine the most likely reactions of these fragments.

III. Reactions of the Primary Fragments

We have listed in part 2 of Table I what we consider to be the most likely reactions of the primary fragments. We have leaned heavily on the current literature regarding the photolytic and pyrolytic decompositions of silane for much of the mechanism as these are currently much better defined chemical systems than that of the discharge plasma.

Reactions (1) through (3) of Table I define the fates of the SiH_2^+ and SiH_3^+ ions generated in the discharge. The work of Lampe and Schmitt (1969) with the γ-ray radiolysis of silane and that of Yu et al. (1972) with the ion–molecule reactions of silane suggest that SiH_3^+ is the precursor of the larger ionic clusters, $Si_xH_y^+$ ($x = 2-5$), commonly observed in the discharge. Reactions (1) and (2) probably proceed via an activated complex, $Si_2H_7^+*$, which can be collisionally stabilized ($k = 2 \times 10^{-26}$ cm^6 molecules^{-2} sec^{-1}), decompose via reaction (2) to form $Si_2H_5^+$ and H_2 ($k = 7 \times 10^{-12}$ cm^3 molecule^{-1} sec^{-1}), or back-react to form SiH_3^+ and SiH_4 (Yu et al., 1972). Lampe and Schmitt's (1969) model suggests that the back reaction is much more likely than either of the two forward reactions, however. Hence the steady state for SiH_3^+ and higher ionic clusters is achieved rapidly and one expects the major loss process for SiH_3^+ to be wall recombination, giving rise to SiH_3. The rate constant for reaction (3) of Table I, which defines the role of SiH_2^+ in the discharge chemistry, is 2.5×10^{-10} cm^3 molecule^{-1} sec^{-1} (Yu et al., 1972). This rapid reaction leads to the production of SiH_3^+ and SiH_3. Having already discussed the reactions of SiH_3^+, we conclude that the major ion–molecule reactions in the discharge lead to the formation of the larger ionic clusters and SiH_3.

Reactions (4) and (9) of Table I define the most likely reactions of the diradical SiH_2. Reaction (4) is an insertion reaction, the measured rate constant being large (John and Purnell, 1973). Thus this reaction is quite rapid and probably adequately represents the almost exclusive fate of SiH_2 in the discharge. Again the insertion of SiH_2 into SiH_4 probably leads to an activated complex, $Si_2H_6^*$, which can be collisionally stabilized to yield disilane or decompose to yield SiH_3SiH and H_2, the products of reaction (9).

The energy barrier for the formation of $Si_2H_6^*$ from SiH_2 and SiH_4 is only 1.3 kcal mol^{-1} (John and Purnell, 1973); hence one would expect collisional stabilization to be more likely than reaction (9). Thus one concludes that the major gas-phase reaction of SiH_2 leads to the formation of Si_2H_6. The SiH_3SiH fragment generated by reaction (9) similarly can insert into SiH_4 [reaction (11)] to yield trisilane, Si_3H_8. Again, though, this probably proceeds through an activated complex that can be stabilized collisionally or decompose to generate a larger diradical fragment. Insertion of this fragment into silane will produce tetrasilane, which either can be stabilized or will decompose in a manner similar to that of trisilane. This sequence of events will lead to the production of pentasilane, pentasilane decomposition will lead to hexasilane, and so on to higher and higher silanes. In fact, however, disilane and trisilane appear to be the only stable higher silanes that occur in significant amounts in the discharge (Longeway et al., 1983). One concludes, therefore, that collisional stabilization is more effective than further decomposition. It has been suggested that this "chain insertion reaction" is the mechanism by which the "powders" commonly observed during silane decomposition are formed.

The reactions of the monoradical SiH_3 are listed as reactions (5), (7), and (8) of Table I. Reactions (5) and (8), respectively, lead to disilane and the silylsilylene (SiH_3SiH) fragment. These reactions are similar to the reactions (4) and (9) of SiH_2, so we will not discuss them further. It is important to note, however, that the energy barrier for creating $Si_2H_6^*$ from $2SiH_3$ is about 25 kcal mol^{-1} (Perkins et al., 1979). Thus one would expect decomposition of this activated complex to SiH_3SiH to play a much more important role than that for SiH_2. Reaction (7) deserves a little more attention. In this reaction, $2SiH_3$ gives rise to SiH_2 and SiH_4, probably through the activated complex mentioned earlier. The rate constant for this reaction has not been measured, but it may represent a significant source of SiH_2 in the plasma if SiH_3 is generated in large enough amounts for the self-collision cross section of SiH_3 to be high. We will say more about this reaction later.

The reaction of H with silane is given by reaction (6) of Table I. The rate constant for this reaction has been measured to be 4.3×10^{-13} cm^3 sec^{-1} (Austin and Lampe, 1977); hence it is also quite rapid, giving rise to SiH_3.

Reactions (12) and (13) of Table I are proposed based on energetic considerations alone. There is little known about the reactions of Si and SiH with silane. The monoradical SiH_3SiH_2 made from reaction (12) can react with silane to abstract H, generating Si_2H_6 and SiH_3, or $2SiH_3SiH_2$ can yield tetrasilane, Si_4H_{10}. The cross section for the latter process is expectedly quite low.

We have summarized the various fates of the primary intermediates in Fig. 2. While the complexity of the discharge chemistry prohibits a complete

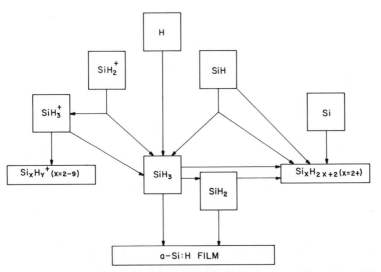

FIG. 2. Schematic diagram outlining the intermediates and products produced by the further reactions of the primary fragments, according to the reactions listed in Table I. The primary fragments are shown in squares. Branched arrows indicate a reaction of the primary fragment that yields more than one product. The production of hydrogen is not shown.

picture of the reaction mechanism, we feel this presentation is adequate. It is our contention that the most important intermediate for film formation in the discharge is SiH_3. We will discuss the basis for this contention in a later section of this paper.

IV. Secondary Decompositions

As the primary decompositions proceed, it is clear that stable reaction products such as disilane and hydrogen will accumulate in the reaction chamber and themselves will undergo electron impact and decomposition. While reaction kinetics alone dictate that the time to achieve steady state will be short (i.e., less than a second), the physical arrangements of the deposition system may impose longer time constants on the system dynamics. For example, in a system wherein a continuous flow of silane is injected into the plasma and chamber gases are pumped from the chamber, the time constants associated with the pumping process will be much longer than any reaction kinetic consideration. This is particularly true for amorphous Si : H deposition, since the stoichiometry of the process requires that more molecules of hydrogen are produced than molecules of silane depleted. Thus one observes that the pressure of the reaction chamber under constant flow and pumping conditions rises during the first several seconds of the discharge. During this time the gas composition of the plasma is changing, and secondary decompositions increase in frequency. Whether such decompositions play an important role in changing the plasma characteristics (e.g., electron density) and the film properties (e.g., hydrogen content of the film) remains to be seen. Nonetheless, we can make a few remarks about what one might expect.

The actual fraction of the gas in the discharge region that is composed of the reaction products, primarily disilane and hydrogen, will be a function of both the flow rate of silane into the system and the discharge current (Longeway et al., 1984). The studies of Longeway et al. (1983) show that the depletion rate of silane at time zero is a linear function of the discharge current. Further studies (Longeway et al., 1984) of the silane discharge under conditions during which the gas is flowing through the reaction chamber and the deposition surfaces are heated show that the depletion rate and product yields remain linear with current over a wide range of discharge currents and are independent of the flow rate. While the depletion rate of silane was unaffected by the surface temperature over the range $50-250°C$, the fractional yields of the higher silanes increased with increasing surface temperature. This indicates that at least a portion of these higher silanes are produced by a surface-catalyzed reaction having a negative activation energy. This result points out the importance of understanding the nature of the surface reactions leading to film formation in that it suggests that species reaching the surface have other reaction pathways available to them that lead to stable gas-phase products. A better understanding of these competitive reaction mechanisms may lead to the discovery of methods for optimizing the film yield. The absence of flow-rate dependence for the product yields suggests that the principal effect of flow rate is to define the residence time of species in the chamber and hence the fractional depletion rate of silane in the chamber. Thus at low flow rates (long residence times) the concentration of reaction products in the discharge will be relatively large, depending on the overall geometry of the reaction chamber.

The electron impact decomposition of H_2 gives rise to H^+, H_2^+, H_3^+, and H atoms. The reaction of H with silane has been discussed [see reaction (6), Table I]. Hydrogen has an ionization potential of 356 kcal mol^{-1} (15.5 eV), as compared with 274 kcal mol^{-1} (12 eV) for silane. Thus one would not expect the decomposition of H_2 to be extensive in a largely silane discharge and hence it would have little effect on the discharge chemistry. Of more importance may be the reactions of H_2 with the primary fragments, but it is beyond the scope of this paper to discuss these reactions at this point.

The decomposition of disilane may be more important than that of hydrogen. Amorphous Si:H films grown from disilane have a higher H content than those grown from monosilane at the same temperature (Catalano, 1983). Since the H content of the film has significant effects on the optical and electrical properties of the film, the accumulation of disilane in the discharge may have serious implications for film properties. This is another question that remains to be answered.

V. Film-Forming Intermediates

Virtually the entire list of primary fragments developed from the previous discussion have been proposed as film precursors. We shall briefly present the evidence for each and reach some tentative conclusions.

Haller (1983) is probably the chief proponent of film growth by an ionic mechanism. The argument, essentially, is that long chain ionic clusters form rapidly in the discharge and play a more important role in silicon transport to wall surfaces than the smaller fragments formed by the primary process. Positive ion impact with the cathode is, of course, an important process in sustaining the discharge, as it acts as a source of secondary electron emission. However, given the very fast reaction rates of ion–molecule reactions in the gas phase (Yu *et al.*, 1972) and the low measured steady-state concentration of ions in the silane plasma (Weakliem, 1980), we are led to conclude that the predominant role of the ions generated in the plasma is the further generation of neutrals [e.g., via reactions (1)–(3) of Table II].

The Si atom has been proposed as the principal film intermediate by Tachibana *et al.* (1982) based on the concentration of this species as deduced from OES measurements and the measured diffusion constant for Si.

Hirose *et al.* (1981) have suggested SiH as the film precursor on the grounds that they measure a direct correlation between deposition rate and the SiH* optical emission intensity. However, Perrin and Schmitt (1982)

TABLE II

RELATIVE INITIAL FORMATION RATES OF H_2,
Si_2H_6, AND Si_3H_8, NORMALIZED TO THE
DEPLETION OF SILANE, $R(x)/R(-SiH_4)$

Compound (x)	$R(x)/R(-SiH_4)$
SiH_4	1.0
H_2	0.69
Si_2H_6	0.39
Si_3H_8	0.03
Solid[a]	0.13

[a] Calculated from Si–atom balance.

have recently measured the emission cross sections for fragments produced by the electron impact with silane and conclude that only 2% of the dissociation process proceeds to SiH*. Additionally, they conclude that there is a closer correlation between the partial pressure of the silane and the emission intensity of SiH* than between the deposition rate and SiH* intensity. This is in agreement with the recent study of Longeway *et al.* (1984) using both mass spectrometric (MS) and OES techniques.

The diradical SiH_2 has been suggested as the film precursor for silane discharges by analogy with CVD deposition systems (Scott *et al.*, 1981). The thermal decomposition of silane is known to proceed to SiH_2 (Neudorfl *et al.*, 1980); hence the film growth mechanism for CVD systems must be via SiH_2. This view is particularly appealing given the rapid insertion rate of SiH_2 into the Si–H bond, which insertion must also be occurring on the growing film surface. We also note that the IR-laser decomposition of silane also proceeds to SiH_2 (Lampe and Longeway, 1981), and recently, amorphous Si:H materials have been grown using an IR laser as the source of silane decomposition (Hanabusa *et al.*, 1979). The only direct evidence that SiH_2 may be the primary intermediate for film growth in the plasma environment comes from Bourquard *et al.* (1981). These researchers compared the fragmentation pattern of SiH_4 in a mass spectrometer with no discharge with the observed fragmentation pattern during discharge. Corrected differences in peak intensities between the two patterns led them to conclude that SiH_2 was present in the discharge to a greater extent than any other radical.

Similar arguments have been marshaled for invoking SiH_3 as the film-forming intermediate. Saitoh *et al.* (1983) have grown a-Si:H material by the Hg-photosensitized decomposition of silane, a technique known to proceed to SiH_3 (Kamaratos and Lampe, 1970). Additionally, Gallagher and Scott (1983) have reported that they have measured mass spectrometrically the flux of radicals to a wall surface of the discharge region. Based on MS appearance potentials, they conclude the SiH_3 is the most abundant radical reaching the film surface.

In the next section we shall outline briefly our own research into the silane discharge chemistry, the conclusion of which is that SiH_3 appears to be the most likely candidate as the film precursor.

VI. Static Pressure dc Silane Discharge Chemistry

We have investigated the chemistry of a static pressure silane discharge (Longeway, 1982) using a combination of two techniques. First, in the static pressure reactor, we have measured mass spectrometrically the initial rates of depletion of silane and formation of the stable reaction products, H_2,

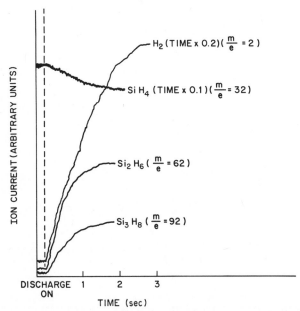

FIG. 3. Mass spectrometric ion current-versus-time tracings demonstrating the production of stable reaction products and the depletion of silane from a silane discharge in a static pressure reaction chamber.

Si_2H_6, and Si_3H_8. The initial rates were determined by monitoring a value of m/e characteristic of the species being measured and calculating the initial slope of the ion current-versus-time tracing as the discharge is turned on. Typical tracings are shown in Fig. 3. The rates in molecules per second were calculated from the measured sensitivity of the MS in amperes per molecule for the species under investigation. The second technique used in this study was to measure the effects on these rates resulting from the introduction of nitric oxide, NO, to the silane discharge. Nitric oxide is a known free-radical scavenger that is highly effective in intercepting SiH_3 but inert toward singlet-state SiH_2 (Perkins $et\ al.$, 1979). These measurements permitted assignment of the relative importance of SiH_2 versus SiH_3 reaction mechanisms in the discharge.

The relative rates of formation of H_2, Si_2H_6, and Si_3H_8, as well as the film, are listed in Table II, normalized to the depletion rate of silane. The depletion rate (at time zero) was found to be independent of silane pressure in the range 0.25–0.75 Torr and linear with current over the range 10–80 mA.

The results of the NO experiments were quite dramatic and are presented in Fig. 4. In the presence of NO the formation rate of disilane was reduced to

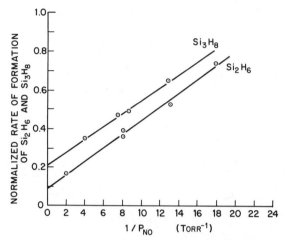

FIG. 4. Reduction in the normalized initial rates of formation of disilane and trisilane (see Table II) in a static pressure silane discharge with the introduction of nitric oxide.

9% of its value in the absence of NO and the formation rate of trisilane was similarly reduced to 21% of its previous value. Not shown in Fig. 4 is the effect of NO on the film formation rate. This value was essentially reduced to zero; i.e., amorphous Si : H formation was not observed in the presence of NO. The results of these studies suggest that the ratio of SiH_3 generation processes to SiH_2 generation processes is 9 : 1.

While it is beyond the scope of this paper to present the details of our analysis, we constructed a mechanism similar to that of Table I and developed a kinetic treatment of the mechanism that expressed the expected yields of the products as well as the relationships between some of the rate constants. This treatment yielded results that were consistent with the observed results and the known rate constants for some of the processes. We refer the reader to the publication of these studies for more details (Longeway, et al., 1983).

It is our conclusion that SiH_3 generation processes dominate in the discharge. We refer to Fig. 2 as to why we feel this conclusion is warranted.

The study is not conclusive, however, because it leaves a major question unanswered — the importance of the disproportionation reaction of $2SiH_3$ to SiH_2 and silane [reaction (7), Table I]. It is possible that this reaction proceeds to a significant extent, leading to much higher concentrations of SiH_2 in the discharge than are produced by the primary processes. Products such as the film and disilane would still be scavengeable by NO under these circumstances, since the precursor to the SiH_2 is now SiH_3. This question must be answered, in part, by further direct measurements of the radical flux to growing film surfaces.

VII. Implications of SiH$_3$ for the Growth Mechanism

Again, it is beyond the scope of this paper to discuss in detail a growth model. However, we feel that a few pertinent remarks are in order as to what one might expect if, in fact, SiH$_3$ is the direct film precursor.

Any growth mechanism must consist of at least four processes to explain the observed film properties. These are

(a) adsorption and/or chemisorption of the film precursor(s) onto the growing film surface,

(b) bond making with the film and/or active site creation,

(c) elimination of hydrogen from the film, and

(d) cross-linking of the film structure.

It is generally observed that the film growth rate is independent of the film substrate temperature. From this one is led to conclude that process (a) probably has a low energy of activation, as well as a rapid reaction rate. However, the hydrogen content of the film is dependent on the substrate temperature. Hence processes (c) and (d) must have a higher energy of activation and may represent rate-limiting steps in the formation of good-quality material.

The silyl radical, SiH$_3$, is known to be quite efficient in the abstraction of hydrogen from other silicon hydrides. We propose, then, that one major step in the film formation process is the creation of dangling-bond sites by H-abstraction from the film surface. Such sites may be active in a number of ways. For example, they can act as bonding sites for other incoming species, or they can act as cross-linking sites within the film. Additionally, such sites may be covered over by the growing film and become incorporated as defect sites in the film structure. Scott *et al.* (1983) has observed that the spin density of plasma-prepared films as measured by electron spin resonance (ESR) is much higher at substrate temperatures below 250°C than the spin density of HOMOCVD films. This is consistent with the SiH$_2$ growth mechanism proposed for CVD materials, as this precursor would most likely undergo insertion at the film surface rather than the creation of dangling bonds. It also suggests, however, since the spin densities of the two materials become similar around 250°C, that temperature-dependent processes such as hydrogen elimination and cross-linking can effectively anneal such sites in the plasma films.

It is also possible that SiH$_3$SiH, a diradical, may act to create dangling bond sites in the film. The singlet state of this fragment is convertible to the triplet state, $\cdot SiH_2SiH_2\cdot$, as per reaction 10 of Table I. The energy barrier for this reaction is only 11 kcal mol^{-1} (Snyder and Wasserman, 1979). One might expect this triplet state to act as a "double monoradical" in the film surface reactions. Such a mechanism may play a significant role in CVD processes as well as plasma processes, since disilane is present in the CVD reactor and its thermal decomposition can lead to SiH_3SiH.

VIII. Future Studies

This brief discussion has probably raised more questions than it has answered. These questions, of course, suggest further avenues of investigation into the discharge chemistry of silane and the film growth process. We list a few such questions here for your consideration.

What is the actual flux of radicals onto the growing film surface? Is SiH_3 more abundant than SiH_2?

Is the disproportionation of SiH_3 to SiH_2 a significant process in the plasma chemistry, acting as a source of higher concentrations of SiH_2?

What are the cross sections for some of the electron impact processes? Does vibrational excitation play a significant role in the decomposition of silane?

What are the details of the growth mechanism?

What effects do secondary decompositions have on the film properties?

What role, if any, do ions play in the growth mechanism? Are they just an additional source of neutral fragments or do they indeed act as film precursors?

These are just a few of the questions that must be addressed in order to gain a better understanding of the discharge chemistry of silane.

REFERENCES

Austin, E. R., and Lampe, F. W. (1977). *J. Phys. Chem.* **81**, 1134.
Bourquard, S., Erni, D., and Mayor, J.-M. (1981). *In* "Proceedings of the Fifth International Symposium on Plasma Chemistry" (B. Waldie and G. A. Farnell, eds.), p. 664. Heriot-Watt University, Edinburgh, Scotland.
Catalano, A. (1983). *Appl. Phys. Lett.* (submitted for publication).
Drevillon, B., Huc, J., Lloret, A., Perrin, J., de Rosny, G., and Schmitt, J. P. M. (1981). *In* "Proceedings of the Fifth International Symposium on Plasma Chemistry" (B. Waldie and G. A. Farnell, eds.), p. 634. Heriot-Watt University, Edinburgh, Scotland.
Gallagher, A., and Scott, J. (1983). Annual report to SERI for contracts Nos. XB-2-02085-1 and DB-2-02189-1, Solar Energy Research Institute, Golden, Colorado.
Griffith, R. W., and Kampas, F. J. (1981). *J. Appl. Phys.* **52**, 1285.
Haller, I. (1983). *J. Vac. Sci. Technol.* **A1**, 1376.
Hanabusa, M., Namiki, A., and Yoshihara, K. (1979). *Appl. Phys. Lett.* **35**, 626.

Hirose, M., Hamasaki, T., Mishima, Y., Kurata, H., and Osaka, Y. (1981). *In* "Tetrahedrally Bonded Amorphous Semiconductors" (R. A. Street, D. K. Biegelsen, and J. C. Knights, eds.), p. 10. Am. Inst. Phys., New York.
John, P., and Purnell, J. H. (1973). *J. Chem. Soc., Faraday Trans. 1* **69**, 1455.
Kamaratos, E., and Lampe, F. W. (1970). *J. Phys. Chem.* **74**, 2267.
Lampe, F. W., and Longeway, P. A. (1981). *J. Am. Chem. Soc.* **103**, 6813.
Lampe, F. W., and O'Keefe, J. F. (1983). *Appl. Phys. Lett.* **42**, 217.
Lampe, F. W., and Schmitt, J. F. (1969). *J. Phys. Chem.* **73**, 2706.
Longeway, P. A., Weakliem, H. A., and Estes, R. D. (1983). *J. Phys. Chem.* November.
Longeway, P. A., Weakliem, H. A., and Estes, R. D. (1984). *J. Appl. Phys.* (submitted for publication).
Matsuda, A., and Tanaka, K. (1982). *Thin Solid Films* **92**, 171.
Neudorfl, P., Jodhan, A., and Strausz, O. P. (1980). *J. Phys. Chem.* **84**, 338.
Perkins, G. G. A., Austin, E. R., and Lampe, F. W. (1979). *J. Am. Chem. Soc.* **101**, 1109.
Perrin, J., and Schmitt, J. P. M. (1982). *Chem. Phys.* **67**, 167.
Saitoh, T., Maramatsu, S., and Migitaka, M. (1983). *Appl Phys. Lett.* (submitted for publication).
Scott, B. A., Plecenik, R. M., and Simonyi, E. E. (1981). *Appl. Phys. Lett.* **39**, 73.
Scott, B. A., Reimer, T., and Longeway, P. A. (1983). *J. Appl. Phys.* December.
Snyder, L. C., and Wasserman, Z. R. (1979). *J. Am. Chem. Soc.* **101**, 5222.
Tachibana, K., Tadokoro, H., Harima, H., and Urano, Y. (1982). *J. Phys. D* **15**, 177.
Weakliem, H. A. (1980). *AIAA Pap.* **80-1327**.
Yu, T.-Y., Cheng, T. M. H., Kempter, V., and Lampe, F. W. (1972). *J. Phys. Chem.* **76**, 3321.

CHAPTER 10

Diagnostics of Silane Glow Discharges Using Probes and Mass Spectroscopy

Herbert A. Weakliem

RCA LABORATORIES
PRINCETON, NEW JERSEY

I. Introduction

Glow discharges are used to deposit a-Si : H films either by the direct plasma decomposition of silane or by reactive sputtering of a Si target using a glow discharge containing some hydrogen in the gas. The gaseous atmosphere of an operating silane glow discharge contains energetic electrons, ions, fragments, and condensation products and the various species are not in thermal equilibrium with each other. The discharge thus provides a rich source of reactive species. The various mechanisms by which the gas-phase species lead to film formation on substrates placed in the glow discharge are discussed in Kampas, Chapter 8, and Longeway, Chapter 9, this volume. The plasma properties that are of interest are (1) composition of the gas phase, including ions and neutrals; (2) plasma density; (3) plasma potential and sheath voltages; (4) electron temperature and energy distribution; (5) ion temperature; and (6) impurity effects. The principal diagnostic techniques used to measure these properties are optical spectroscopy, electrical measurements, and mass spectroscopy. The discussion in this chapter is confined to electric probe techniques and mass spectroscopy of glow discharges excited in pure silane and various silane–gas mixtures.

195

II. The Structure of Glow Discharges

1. General Aspects

An extensive literature exists on the theory of glow-discharge phenomena; however, the aspects that are of primary concern here may be found in books by Chapman (1980) and Francis (1960). The glow discharges we are concerned with are typical cold plasmas; the ions and electrons are not in thermal equilibrium with each other. The ion temperature is typically a few hundred degrees Celsius, at most, whereas the electron temperature T_e may range from 1000 to 10,000°K (0.1 – 10 eV). The plasma densities are in the range $10^7 – 10^{12}$ cm^{-3}. There exist regions in the glow discharge that may be described as true plasmas; free charges are present, but overall charge neutrality is maintained ($n_+ \sim n_-$). Space-charge sheaths are established next to surfaces in contact with the glow, which determines the electric fields and thus the flow of charged particles to and from the surfaces.

The mechanism by which the glow discharge is initiated and maintained and the voltages and electric fields that are present depend on the excitation method used, on the composition and geometrical properties of the electrodes, and on the gas in which the discharge is excited. The distinctive characteristics of dc and rf excited discharges are described in the next sections. Descriptions of the various types of reaction chambers and methods of excitation used for the deposition of thin film a-Si:H are discussed in detail in Hirose, Chapter 2, and Uchida, Chapter 3, this volume.

2. dc Discharges

A dc discharge is sustained by the interaction between the cathode and the luminous negative glow. Positive ions from the negative glow are accelerated across the cathode fall, typically a few hundred volts in silane, where they bombard the cathode, producing electron emission. Fast electrons from the cathode produce ionization primarily in the negative glow region. The electron and ion production and the cathode fall voltage are determined by a balance between the current supplied by the external circuit and the various energy loss processes in the plasma — wall recombination, heating, radiation, and so on.

The negative glow is only 1 – 3 cm wide, depending on the silane pressure, and is followed by the Faraday dark space; and if the anode is sufficiently far from the cathode a positive column is present. The positive column is a true plasma region whose properties differ from those of the negative glow. The positive column may contain stationary or moving dark striations, depending on the walls and the flow of gas. Films of a-Si:H may be deposited on the cathode, the anode, or a separate substrate placed behind a screened cathode, the latter being referred to as a proximity electrode assembly (Carlson,

1977). The sheath voltages at the anode and the proximity electrode are less than that at the cathode.

3. RF DISCHARGES

a. General Considerations

We consider the discharge to be rf excited when the excitation frequency is greater than a few hundred kilohertz and thus the heavy positive ions cannot instantaneously follow the field. The most commonly used frequency is 13.56 MHz, which is alloted by international agreement for "industrial, scientific, and medical" (ISM) purposes (Vossen and Cuomo, 1978). For excitation frequencies ω and pressures such that $\omega \gg \nu_c$, where ν_c is the collision frequency for ionization, an additional ionization mechanism becomes important. The electron, which does instantaneously respond to the rf field, may extract energy from the rf field via inelastic collisions with gas molecules in the bulk, thereby gaining sufficient energy to ionize the gas. This volume excitation is an additional source of ionization to that caused by secondary electron emission from the electrodes and walls. The maintenance potential of rf discharges is therefore lower than that found for dc discharges, and the sheath voltages are also different for the same pressure and power.

b. Inductively Coupled Discharges

A discharge may be excited by an oscillating current in a solenoid wrapped around the discharge vessel. At very high discharge power a true magnetic coupling occurs, whereby an oscillating current in the gas is induced that produces an oscillating H field opposite to that induced by the current in the external coil. However, the discharges of interest here are relatively low power and the coupling is not magnetic; rather, the excitation is primarily via the electric field that is induced by the surface charge and capacitance of the insulating wall of the chamber. This type of discharge is not widely used in the field of thin-film deposition.

c. Capacitively Coupled Discharges

A pair of capacitor plates, either internal or external to the chamber, to which an rf field is applied is the most common method used to excite an rf discharge, referred to here as rf(C) excitation. Numerous modes exist in the discharge, depending on the frequency ω, the pressure, and the size of the electrodes and the shape of the container. A chamber having a symmetric pair of electrodes, with one serving as the substrate for the film growth, is in common use (Knights, 1976). A pair of asymmetric semicylindrical electrodes mounted external to a cylindrical glass container and having a heated substrate whose surface is perpendicular to the cylinder axis has also been used (Okamoto *et al.,* 1979). Chambers designed for sputtering application have also been successfully employed (Hanak and Korsun, 1978; Weakliem, 1980).

The sputtering type of chamber often employs a metal chamber that serves as one electrode, and a second electrode, usually much smaller than the chamber wall, is capacitively coupled to the rf power supply (Vossen and Cuomo, 1978). This is referred to as an asymmetric diode and has the property that a negative dc self-bias is developed on the smaller electrode (called the cathode or the target, in sputtering terminology) whose magnitude is related to the relative areas of the two electrodes. We also note that all surfaces in contact with the glow are more negative than the average dc potential of the plasma and thus are subject to positive ion bombardment. The potential of the substrate may be controlled externally and several studies have investigated the effect of external bias on properties of the plasma and the deposited film (Knights *et al.,* 1979; Hotta *et al.,* 1981; Matsuda and Tanaka, 1982).

The glow of a rf(C) discharge, particularly one operated at power and pressures enabling the glow to fill the entire chamber volume, is a plasma whose features are well described by the theory of positive columns (Francis, 1960). At higher pressures (greater than about 0.1 Torr), particularly in asymmetric diode systems, the glow becomes attached to the cathode and may become quite nonuniform and the dc potential of the target becomes quite small. The power absorbed by the gas is usually taken to be the difference between the forward and reflected power as measured by a diode. It has been shown that substantial nonlinearities exist in the current waveforms and unless care is taken to match the discharge properly to the external power source, as much as 75% of the power may be dissipated in eddy currents via the leaky capacitor of the discharge chamber (Mosburg *et al.,* 1982).

4. MAGNETIC CONFINEMENT

The plasma of the glow discharges may be confined by the application of magnetic fields. The magnetic field is oriented so that an electron drift component is induced perpendicular to the $E \times B$ plane, which tends to confine the electron to the region adjacent to the electrode to which the magnetic field is applied. The increased residence time of the electron in the vicinity of the electrode increases the ionizing efficiency. The primary plasma region in which the electrons are confined is an intense glow having characteristics similar to the negative glow of a dc discharge, and a positive columnlike region extends to the counter electrode. The ions, because of their large mass, are not confined to the primary plasma region by the weak magnetic fields which are used.

Magnets attached to a planar electrode of an rf(C) discharge convert the electrode to a planar magnetron (Waits, 1978) that has been used for the deposition of a-Si : H films (Weakliem, 1980). In this case, the primary plasma region of the planar magnetron is a luminous ring lying parallel to the surface of the electrode. Magnetic confinement is also used in a multipole discharge, where electrons from hot filaments are accelerated by a dc potential to high energy, thereby initiating and maintaining the discharge. Magnets attached to the wall of the chamber confine the discharge near the wall, and this system has been used for the deposition of a-Si : H and simultaneous study of plasma parameters (Drevillon et al., 1980).

Magnetic confinement was also used to study the effects on film growth in a system employing rf excitation via external ring electrodes in a cylindrical discharge chamber (Taniguchi et al., 1980; Hamasaki et al., 1980). It was found that the magnetic field affected the emission intensities of SiH, H, and H_2 in the glow and also influenced the distribution of hydrogen in the film between sites characterized by monohydride (Si – H) and dihydride (Si – H_2) bonding groups. Low-temperature crystallization of the film was also found for a phosphine – silane rf excited, magnetically confined, glow discharge used to deposit phosphorus-doped silicon – hydride films (Hamasaki et al., 1980).

III. Probe Measurements

The use of probes to measure plasma characteristics is a relatively simple experimental technique, but the results may be difficult to interpret (Chen, 1965). The probe usually consists of an insulated wire having a short section of the tip exposed, or it may be a plane surface in contact with the plasma. A simplified circuit diagram, shown in Fig. 1, illustrates the essential features of the technique. The probe is biased by a power supply or battery and the probe current is measured with respect to a reference electrode in contact with the plasma. The resulting current – voltage characteristic may be analyzed in favorable circumstances to give the plasma electron temperature T_e, the plasma potential V_p, and the plasma density n_e.

Although the measurement is straightforward, the results may be strongly influenced by probe – plasma interactions. For the plasmas of interest here, the deposition of a film on the probe, with a concomitant alteration of its surface conductivity and charge sheath, is the most serious one. It has been reported, however, that at least for pure silane discharges, the deposition of a-Si : H film on the probe does not affect the measured $I - V$ characteristic (Perrin et al., 1982; Mosburg et al., 1983). Apparently, the conductivity of the film is sufficiently high that the surface potential is not affected by the presence of the film and thus the plasma sheath is not affected. In the case of a discharge in $SiH_4 - B_2H_6$, the deposited film does affect the results and

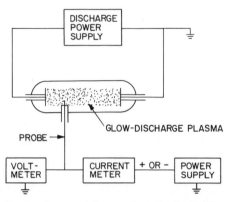

FIG. 1. Schematic showing the essential parts of a probe circuit. The probe $I-V$ characteristic is measured with respect to one of the electrodes in contact with the glow, indicated as a circuit common point.

reliable probe measurements could not be made (Mosburg *et al.*, 1983). The use of a heated (emissive) probe to minimize surface contamination has been reported (Perrin and Schmitt, 1982); however, it was needed only to clean the probe of contaminants initially present. After the initial cleaning, the characteristics obtained in silane discharges were found to be quite stable. A screened probe, whose intended use was for the determination of the positive ion flux, was found to have a stable characteristic that could be analyzed to yield n_e, T_e, and V_p (Weakliem, 1982). In cases where the deposited film does create problems, the use of pulsed potential and current collection together with a movable shield to protect the probe surface between measurements could be employed.

The probe characteristic for positive bias is shown in Fig. 2 for a dc discharge in SiH_4 where a screened probe was placed near the edge of the negative glow. Electrons are collected by the probe for a positive bias, and if the electrons have a Maxwell distribution with a temperature T_e, the electron current is given by

$$I_e = \frac{en_0\bar{v}_e}{4} A \exp\left[-\frac{e(V_p - V)}{kT_e}\right], \tag{1}$$

where $\bar{v}_e$ is the random electron speed, A is the effective collection area of the probe, and V is the applied potential. A plot of $\ln I_e$ versus V should give a straight line having a slope e/kT_e. Experimentally, the net current, ions and electrons, is measured; thus the measured current must be corrected by subtraction of the ion current in order to obtain T_e.

A simple Maxwell distribution was not found for either the multipole discharge or the dc discharge. In both cases the data were interpreted as

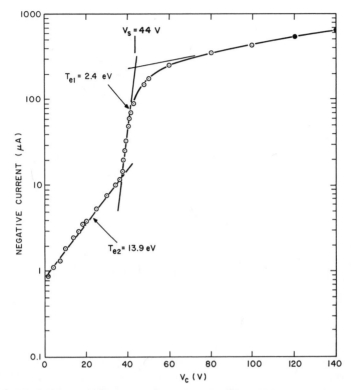

FIG. 2. Measured negative current versus probe voltage characteristic measured at the edge of the negative glow in a dc silane discharge using a screened probe. $p = 0.48$ Torr. 90 sccm silane.

comprising contributions from two groups of electrons having different temperatures. The multipole discharge had a low-energy group with $T_e \sim 1$ eV, whereas the high-energy group had an energy close to eV_c, where V_c was the potential of the electron-emitting cathode, as shown in Fig. 3. This result shows that a substantial number of electrons emitted from the cathode suffer no collisions during passage through the cathode dark space. The plasma densities reported range from 1 to 10×10^9 cm^{-3}.

The low-energy group of electrons results from the numerous low-energy inelastic electron–molecule collisions that cause molecular excitation, including electron impact dissociation of silane to produce neutral SiH$_n$ ($n = 0-3$) fragments, H, and H$_2$. The energies of these reactions range from 2 to 5 eV (Kocian, 1980; Turban et al., 1982).

Since a hot filament source was not used for the dc discharge (Weakliem, 1982), the higher-energy group of electrons may represent that fraction of

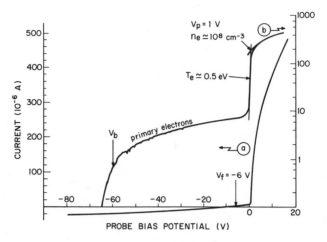

FIG. 3. Typical Langmuir probe characteristic for a silane plasma operated at $p = 1$ mTorr, and a filament bias of 60 V in a multipole reactor: (a) as measured, (b) after subtraction of the ion current. The arrow associated with each plasma parameter indicates which part of the curve contributes to its determination. (From Perrin et al., 1982.)

the electrons emitted by the cathode that traverse the negative glow essentially without collision. The relative fraction of the higher-energy group was found to increase as the silane flow decreased. It should also be noted that there is a substantial increase in the hydrogen content of the plasma gas mixture as the silane flow decreases for constant power dissipated in the discharge. The plasma densities were determined from the measured positive ion flux densities and n_e was found to lie in the range 2×10^7 cm^{-3} to 1×10^8 cm^{-3}.

The dependence of T_e and E/p (the reduced electric field) on pressure was obtained from probe measurements in SiH$_4$ and SiH$_4$–Ar mixtures (Kocian et al., 1979). It was shown that at fixed pressure E/p is larger and T_e is smaller for a pure SiH$_4$ discharge compared with a SiH$_4$–Ar discharge. This was explained to be a result of the existence of numerous low-energy excitation processes available in the molecule silane as compared with the high-energy processes available to the atom Ar. Since the excitation processes of hydrogen molecules have much larger energies than those of silane, this may partially account for the increasing contribution of high-energy electrons as the hydrogen content increases. The minimum energy for the electron impact dissociation of hydrogen, reaction (I), is $E_d \sim 8.5$ eV (Goodyear and von Engel, 1962):

$$e + H_2 \rightarrow 2H(1s) + e. \tag{I}$$

Measurements of the plasma density in a 5% SiH_4 in He discharge excited at 13.56 MHz via two ring electrodes using microwave interferometry (10 GHz) have been reported (Turban *et al.*, 1979). This method is excellent for directly obtaining the average electron density, particularly for an extended nearly uniform discharge. The plasma density was found to have a square root dependence on the rf power, $n_e = bW^{1/2}$, where W is the rf power. The measured values of n_e ranged from 3×10^9 cm^{-3} to 1.8×10^{10} cm^{-3} for rf power ranging from 10 to 250 W.

A probelike characteristic was also obtained from measurements of the dc bias potential applied to large-area substrates in a discharge chamber excited at 13.56 MHz using external electrodes (Hotta *et al.*, 1981). The dc field was applied perpendicular to the rf field. Values for the plasma potential of 60 and 110 V were deduced from the I–V characteristic for SiH_4–H_2 discharge operated at 35 W rf and 55 W rf, respectively.

Although knowledge of the plasma parameters and sheath voltages would be quite useful for modeling the discharge, it is apparent that the data are sparse and fragmentary. The application of probe theory results to some of the measurements is certainly open to question, so the results must be used with caution. In almost all cases a Maxwellian electron distribution, or two groups of electrons having different Maxwell distributions was assumed; however, the actual distribution might be quite different, and future measurements should be aimed at directly measuring the distribution.

IV. Mass Spectroscopy

5. SAMPLING THE DISCHARGE

Mass spectroscopic analysis of the gas composition of a glow discharge is capable of providing detailed information about the chemical reactions that occur in the discharge. Knowledge of the identity and the energy of the species arriving at a substrate on which a thin film of a-Si : H is deposited is of fundamental importance. The technique has been shown to be quite valuable in studies concerning sputtering applications (Coburn and Kay, 1972; Purdes *et al.*, 1977).

The need to analyze the discharge without undue disturbance of the plasma is one of the principal difficulties. The discharge must be sampled through a suitable aperture in order to reduce the pressure from that of the deposition chamber, P_D, which typically lies in the range 1×10^{-3} to 1 Torr, to a pressure sufficiently low to allow safe operation of the mass spectrometer (less than 1×10^{-5} Torr). The pressure in the mass spectrometer chamber P_M is determined by the area of the aperture A and the effective pumping speed, S_{eff} at the mass spectrometer side of the aperture according to

$$P_M = P_D \, \bar{v} A / (4 S_{eff}).$$ (2)

We have assumed the aperture to be circular and $\bar{v}$ is the mean molecular speed in the deposition chamber. The requirement that the aperture should not unduly perturb the plasma in its vicinity means that its diameter should be less than the sheath thickness d (Hasted, 1975). If the aperture is considered to be a probe at a potential V with respect to the plasma, the current density j at the orifice is given by the Child–Langmuir law:

$$j = \frac{1}{9\pi} \frac{2e^{1/2}}{M} \frac{V^{3/2}}{d^2}.$$ (3)

The units in Eq. (3) are cgs and M is the mass of the charged particle. The total ion current collected through the aperture whose area is A is $I = jA$.

The values of j and V are not generally known, and Eq. (3) is only an approximation valid for collisionless extraction, so d is difficult to estimate. If the aperture is too large, causing the sheath to collapse, the plasma may penetrate the low-pressure region behind the aperture and edge effects could strongly influence the electric fields. The mass spectrum measured in this instance would not necessarily represent the state of affairs present in the undisturbed plasma. Although the aperture should be made as small as possible, experience has shown that apertures having diameters ranging from 50 to 300 μm appear to be satisfactory for sampling silane discharges. The need to avoid plugging the aperture with a solid deposit and to have a sufficiently large ion current appears to give a practical lower limit of about 25 μm diameter.

A single stage of differential pumping may serve to sample the discharge, as shown by the following estimate. If the pressure in the discharge chamber is 1 Torr and the mass spectrometer chamber is to be maintained at a pressure less than 1×10^{-6} Torr, taking $\bar{v} = 4 \times 10^4$ cm sec^{-1} and for an aperture diameter of 50 μm, the required pumping speed calculated using Eq. (2) is $S_{eff} = 200$ liters sec^{-1}. A medium-size diffusion pump or turbomolecular pump could be used in this case.

In order to obtain more flexibility in the choice of aperture size and the operating pressure range, a dual, differentially pumped sampling arrangement is often used (Haller, 1980; Turban et al., 1980; Weakliem, 1980). A schematic diagram showing the essential features of a dual chamber sampling system is shown in Fig. 4. Electrostatic lenses may be mounted in the first chamber in order to extract and focus ions onto the second aperture. If no focusing is used, the distance between the apertures in the first chamber must be kept short to minimize possible ion–molecule collisions so that the

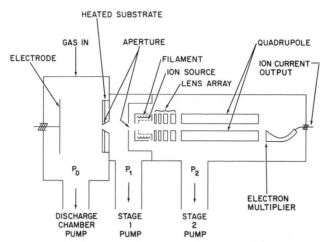

FIG. 4. Schematic representation of a dual differentially pumped chamber for the mass spectroscopic detection of neutrals and ions produced in a glow-discharge chamber.

distribution measured in the second analyzer chamber faithfully represents the distribution in the discharge chamber. The apertures must be aligned with the center line of the quadrupole axis and the focusing lenses contained in the second, analyzer chamber should allow line-of-sight passage of ions from the discharge chamber. The ions from the discharge chamber that are detected are contained in a solid angle determined by the area of the first aperture and its distance from the analyzer lens system in the second chamber.

The ion signal may be substantially increased by using an electrostatic lens system in the first chamber to focus ions from the discharge onto the second aperture of the analyzer. The solid angle that is collected is determined by the lens system used. Since the physical dimensions of the lens system require a distance of 10–20 cm, care must be taken to keep the pressure in the first chamber low enough to minimize ion–molecule collisions. The slow ions are particularly vulnerable to loss by charge exchange or condensation reactions.

Ions that reach the second, analyzer chamber are focused onto the quadrupole entrance by an array of electrostatic lenses and they are mass analyzed by passage along the quadrupole. Neutrals from the discharge chamber may also be analyzed after they are ionized. This is achieved by electron impact ionization by electrons that are emitted from a filament and accelerated to a field-free region called the ion source, which is outlined by dotted lines in Fig. 4. The ions created in the ion source region are then focused onto the quadrupole entrance aperture. The detector may be one of a number of electron multipliers, although the continuous dynode electron multiplier,† which is shown in Fig. 4, has been found to be a long-lived, reliable, sensitive detector for silane ions.

In some instances there may be interference between ions produced in the discharge and ions produced from neutrals in the ion source. This can be observed by turning off the electron source current and measuring the remaining ion current. It might be necessary to deflect the discharge-produced ions magnetically or the sampling orifice could be mounted off-axis. In the case of discharge ion detection, another source of interference may occur, arising from ionization in the analyzer chamber caused by energetic electrons from the discharge that emerge through the sampling aperture. Such interference would be expected for very energetic discharges or when auxiliary electrostatic lenses behind the sampling orifice are positively biased, in order to retard the positive ions, or to analyze their energies.

One must also consider the sheath through which the ions are accelerated before being analyzed. A large voltage drop would promote energetic reactions, and the final composition and energy of the distribution is strongly influenced by charge exchange in the sheath (Davis and Vanderslice, 1963). The effects on the ion composition as a function of sampling through different sheaths of rf discharges excited in methane have been studied (Vasile and Smolinsky, 1975). The differences between the ion compositions found for axial and radial extraction were attributed to the difference in the energies of the electrons and ions in different sheath voltages.

There are numerous mass filtering devices and methods available for mass spectroscopy, including static magnetic deflection and dynamic mass analyzers such as time-of-flight analysis and quadrupole analyzers. All the work described here was done using quadrupole analyzers, and although the results should not depend on the method of mass filtering employed, the mass discrimination characteristics of the filter and detector used must be borne in mind when interpreting the data. Quadrupoles have a mass dependent transmission that depends on the resolution (Austin *et al.*, 1976). Heavy ions tend to be discriminated against; consequently, the observed distribution may seriously underestimate the abundance of the larger, heavier ions.

6. IONS

a. Primary Processes

We will be concerned almost exclusively with positive ions, which play a major role in the establishment and maintenance of the glow discharges.

---

† The Channeltron is a registered trademark of the continuous dynode multiplier manufactured by Galileo Electro-Optics Corp., Sturbridge, Massachusetts.

Negative ions containing up to four silicon atoms have been detected in small concentration in a low-pressure multipole discharge (Drevillon *et al.*, 1981). The mechanism by which the negative ions are formed has not been studied, and since they remain effectively trapped in the plasma region, they affect the deposition process only indirectly. The positive ion reactions occur over a much wider energy range; they play an essential role in initiating and maintaining the discharge, and their energetic bombardment of surfaces may affect the properties of the deposited a-Si:H films. The following discussion is concerned exclusively with positive ions.

The primary ionization processes in a silane glow discharge involve the following four dissociative ionization processes:

$$e + SiH_4 \rightarrow SiH_3^+ + H + 2e \qquad (E_a = 12.3 \text{ eV}), \qquad \text{(II)}$$

$$e + SiH_4 \rightarrow SiH_2^+ + H_2 + 2e \qquad (E_a = 11.9 \text{ eV}), \qquad \text{(III)}$$

$$e + SiH_4 \rightarrow SiH^+ + H_2 + H + 2e \qquad (E_a = 16.1 \text{ eV}), \qquad \text{(IV)}$$

$$e + SiH_4 \rightarrow Si^+ + 2H_2 + 2e \qquad (E_a = 11.7 \text{ eV}). \qquad \text{(V)}$$

The energies given are the appearance potentials (Potzinger and Lampe, 1969), the minimum energy at which the process occurs with no excess energy imparted to the products. The ionization cross section rises steeply from the threshold energy with an energy dependence generally accepted to be linear:

$$Q = K(E - E_a). \qquad (4)$$

The cross section reaches a broad maximum in the neighborhood of 50–100 eV and then falls off as $1/E$ (Mott and Massey, 1965). If the electrons have a distribution $f(E)$ (not necessarily Maxwellian) and the molecular number density is n, the number of electron–ion pairs created per unit time per unit electron Z is

$$Z = n \int_0^\infty Q(E)(2E/M)^{1/2} f(E) \, dE \qquad (5)$$

where M is the molecular mass and the electron distribution is normalized $\int_0^\infty f(E) \, dE = 1$. The dissociation ionization cross sections for energies up to 100 eV are shown in Fig. 5. The primary ions may react with silane molecules to initiate a chain reaction sequence leading to the production of large silicon hydride clusters.

b. Ion–Molecule Reactions in Silane

Ion clusters $Si_x H_y^+$ with $x = 1$–9 and $y \leq 2x + 3$ have been observed in silane glow discharges excited by various means (Gallagher and Scott, 1981;

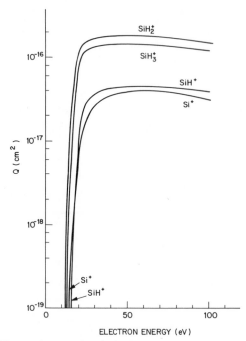

FIG. 5. Electron impact dissociative ionization cross sections of silane.

Haller, 1980; Turban *et al.,* 1980, 1981; Weakliem, 1980). The SiH_3^+ ion is usually found to be the dominant one, but there are exceptions. We shall use I_{xy} to denote the ion current of the $Si_xH_y^+$ion; $I_x = \Sigma_y I_{xy}$ denotes the total ion current of all the ions containing x Si atoms, and the relative abundance $A_x = I_x/\Sigma_x I_x$ is the fraction of ions containing x Si atoms of the total ion composition.

Haller (1980) found that A_x/A_{x-1} was about 0.3, nearly independent of x, pressure and power for an inductively excited 13.56-MHz silane glow discharge. The rf power was 1.3–14 W, p was 0.017–0.240 Torr, and the flow was 0.2–3.4 cm³ min⁻¹ at STP (sccm). A chain reaction mechanism involving the stepwise addition of SiH_4 and hydrogen transfer was proposed to explain the results. In contrast, discharges in silane using both dc (Gallagher and Scott, 1981) and rf capacitive excitation (Weakliem, 1982) showed a marked dependence of the relative abundances A_x on both pressure and discharge power, as shown in Figs. 6–8.

The principal ion–molecule reactions leading to large Si-containing clusters involve the reaction of an ion with silane to increase the size of the

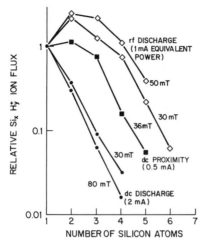

FIG. 6. Relative $Si_xH_y^+$ ion fluxes, I_x/I_1, for three types of low-power discharges in silane. (From Gallagher and Scott, 1981.)

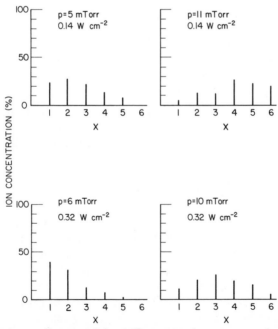

FIG. 7. Abundances of ions containing x silicon atoms shown as percent ion concentrations versus x for rf magnetron discharges operated in pure silane.

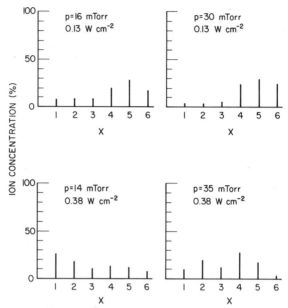

FIG. 8. Abundances of ions containing x silicon atoms shown as percent ion concentration versus x for rf diode discharges operated in pure silane.

ion by one Si atom. The generalized ion–molecule reaction may be written

$$Si_xH_y^+ + SiH_4 \rightarrow Si_{x+1}H_z^+ + mH_2 + nH, \qquad (VI)$$

where $4 + y = 2m + n + z$. The steady-state ionic composition depends on the silane partial pressure at equilibrium and on the reaction cross sections, which are functions of the relative energies of the reactants. These reactions may occur both in the bulk plasma and in the sheath through which the ions are accelerated to reach the sampling aperture. The accelerated ions may also participate in charge exchange and fragmentation reactions.

The sheath voltage at the cathode of a dc discharge is large, as is the sheath voltage at the rf-powered electrode of an asymmetric diode discharge. The magnitude of the dc self-bias potential of a rf(C) discharge is related to the relative sizes of the electrodes (Coburn and Kay, 1972). However, the dc bias on the rf-powered electrode is less than the cathode fall voltage of a dc discharge for the same power dissipation.

Ions sampled at the cathode of a dc discharge have a distribution consisting of energetic, single Si-containing ions largely stripped of hydrogen; the concentrations are $Si^+ > SiH^+ > SiH_2^+ > SiH_3^+$, as shown in Fig. 9 (Weakliem, 1982). The distribution is in the reverse order for ions sampled through a substrate placed near a screened cathode, and these ions have low

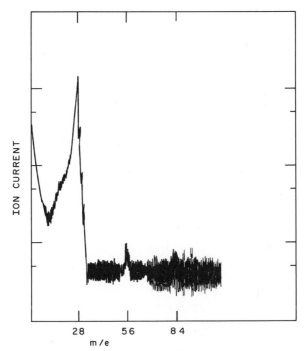

FIG. 9. Relative ion current of a dc discharge in silane in which the ions were sampled through the cathode. The broad low mass tail is characteristic of the spectrum of very energetic ions.

energies. This is shown in Fig. 10, where the sampling orifice is biased successively more negative with respect to the floating potential. The relative abundances of the SiH_y^+ ions were found to depend on substrate bias, where the ions were sampled through the substrate of a rf(C) excited discharge (Matsuda and Tanaka, 1982). The deposition rate was found to be relatively independent of bias; however, the hydrogen incorporation in the film did depend on bias. An increase in the negative bias increased the abundance of H^+ in the plasma and also increased the concentration of hydrogen in the film.

Ions sampled transverse to the electrodes from the negative glow region of a dc discharge also have a distribution whose abundances A_x have a maximum moving to larger x as the pressure increases, as shown in Fig. 11 (Gallagher and Scott, 1981). The ion abundances for rf diode and rf magnetron discharges sampled through the anode (the large-area grounded electrode) also have a distribution in which the size of the cluster increases as the pressure increases and the rf power decreases, as shown in Figs. 7 and 8. The sheath voltage in this case is equal to the plasma potential.

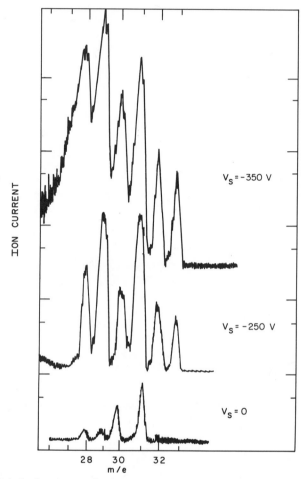

FIG. 10. Relative ion currents for the monosilane ions as a function of bias of the sampling aperture V_S, where $V_S = 0$ is the floating potential. The sampling aperture was in proximity to a screened cathode of a dc silane glow discharge.

The ion distribution for monosilicon hydrides sampled through low-voltage sheaths usually shows that SiH_3^+ is more abundant than SiH_2^+, and both are more abundant than Si^+ and SiH^+. The dominance of the SiH_3^+ ion, except at very low pressure, has been attributed by Turban *et al.* (1982) to the large cross section of the reaction

$$SiH_2^+ + SiH_4 \rightarrow SiH_3^+ + SiH_3, \qquad (VII)$$

where $Q = 230 \times 10^{-16}$ cm^2 (Henis *et al.*, 1972).

The steady state concentrations of the monosilane ions were calculated

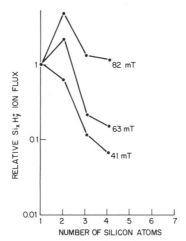

FIG. 11. Relative $Si_xH_y^+$ ion fluxes, I_x/I_1, sampled from the side of the negative glow region of dc silane discharges operated at the pressures indicated. (From Gallagher and Scott, 1981.)

for various pressures and flows, for a cylindrically symmetric reactor by simultaneous solution of the rate equations for the generation of ions by electron impact ionization and ion–molecule reaction and the loss of ions by diffusion and loss by ion–molecule reaction (Turban *et al.*, 1982). The results showed that SiH_3^+ is most abundant for pressures greater than about 0.1 Torr, whereas at very low pressure SiH_2^+ is dominant.

In the case of magnetically confined discharges, the SiH_3^+ ion always appears to be dominant, even for pressures as low as 1 mTorr. The abundance ratio $A_{13}/A_{12} \approx 30$ for a 13.56-MHz magnetron discharge with p in the range 5–15 mTorr (Weakliem, 1980) and $A_{13}/A_{12} \approx 3$ for a dc multipole discharge at $p = 1$ mTorr (Drevillon *et al.*, 1980). (Recall that A_{xy} denotes the abundance of the ion $Si_xH_y^+$.) Since the mean electron energy is reduced by the presence of the magnetic field, which increases the electron collision probability for a given pressure (Francis, 1960), and since the ionization cross section for the production of SiH_2^+ is always larger than that of the production of SiH_3^+, it would seem that a lower electron energy would favor the production of SiH_2^+.

The concentration of SiH_2^+ may be dominant in the primary plasma ring of the magnetically confined discharges. However, if the sampling aperture of the mass spectrometer is somewhat remote from the primary plasma ring, the measured composition may differ from that in the ring. We estimate the mean free path to be of the order a few centimeters for pressures in the range 1–10 mTorr. Therefore, if the sampling aperture is 10 cm or more from the primary plasma ring, as it was in the case of the magnetron discharge

(Weakliem, 1980), the ion–molecule reaction (VII), which produces SiH_3^+ from SiH_2^+, may account for the large A_{13}/A_{12} ratios observed.

The tendency for SiH_3^+ to dominate is further promoted by the fact that the hydrogen exchange reaction (VIII) has a larger cross section than reaction of SiH_3^+ with SiH_4 to produce secondary ions (Yu *et al.*, 1972); and the reaction of hydrogen ions with SiH_4 to produce SiH_3^+ also has a large cross section (Allen *et al.*, 1977):

$$SiH_3^+ + SiH_4 \rightarrow SiH_4 + SiH_3^+, \tag{VIII}$$

$$H_n^+ + SiH_4 \rightarrow SiH_3^+ + H_2 + H_{n-1}. \tag{IX}$$

The principal reaction leading to the depletion of SiH_3^+ appears to be

$$SiH_3^+ + SiH_4 + M \rightarrow Si_2H_7^+ + M. \tag{X}$$

The association product ion $Si_2H_7^+$ is collisionally stabilized by a third body M (which could be another SiH_4 molecule), and the reaction is thus favored by increasing pressure (Yu *et al.*, 1972).

c. Ion–Molecule Reactions in Gas Mixtures

(1) SiH_4-H_2 *and* SiH_4-D_2. It was noted earlier that the addition of hydrogen to a silane glow discharge tends to increase the degree of hydrogenation of the silane ions and increases the abundance of SiH_3^+ via (IX). The higher collisionally stabilized ions $Si_xH_{2x+3}^+$ are also found in increased abundance when the hydrogen content of the discharge is increased. Silane–deuterium discharges also produce SiH_3^+ as the dominant ion, owing to the large cross section of the following reactions:

$$D_2^+ + SiH_4 \rightarrow SiH_3^+ + D_2 + H, \tag{XI}$$

$$D_3^+ + SiH_4 \rightarrow SiH_3^+ + D_2 + HD \tag{XII}$$

(Turban *et al.*, 1981; Allen *et al.*, 1977). The partially deuterated ions SiH_2D^+ and $SiHD_2^+$ are also observed but result from reactions having relatively low cross sections, $Q < 10^{-16}$ cm^2 (Turban *et al.*, 1981).

Hydrogen not only provides additional reaction pathways for the production of silane ions but also has been found to etch a-Si:H solids deposited on the reactor walls (Turban *et al.*, 1980). The ion spectrum of a hydrogen discharge operated in a chamber with freshly deposited a-Si:H on the walls contains the ions SiH_3^+, $Si_2H_y^+$, as well as the expected H_2^+ and H_3^+ ions. The film may also be etched by atomic H (and D in the case of a deuterium discharge); the products SiH_4 and Si_2H_6 are observed in a hydrogen discharge; and SiD_4, Si_2D_6, and HD are found in the gas phase of a deuterium discharge. Silicon chemical transport was also observed in a dc hydrogen discharge in contact with c-Si (Veprěk and Mareček, 1968; Veprěk, 1976), and this process involves the etching of Si by H to produce gaseous SiH_4, which then decomposes on a hot surface to redeposit Si.

(2) *SiH_4 - Rare Gases.* Silane is often diluted in rare gases He, Ne, Ar, or Xe in order to increase the utilization of silane or to minimize the gas phase nucleation that can produce precipitates of a silicon – hydride polymer from the discharge. If silane is the minor constituent, the principal primary ionization proceeds via the rare gas R:

$$e + R \rightarrow R^+ + 2e. \qquad (XIII)$$

The production of silane ions takes place by charge exchange (XIV) or Penning ionization via the rare-gas metastable R* (XV):

$$R^+ + SiH_4 \rightarrow R + SiH_x^+ + \text{hydrogen products}, \qquad (XIV)$$

$$R^* + SiH_4 \rightarrow R + SiH_x^+ + e + \text{hydrogen products}. \qquad (XV)$$

The few ion spectra that have been reported are not markedly different from silane ion spectra (Turban *et al.*, 1979; Gallagher and Scott, 1983). Detailed interpretation of the ion spectra must await measurement of the cross sections of (XIV) and (XV) as a function of relative collision energy. Initial results on the measurements of charge exchange cross sections for He^+, Ne^+, and Ar^+ with SiH_4 and Si_2H_6 have recently been reported (Gallagher and Scott, 1983; Chatham *et al.*, 1983).

(3) *SiH_4 - CH_4.* Glow discharges of 5% (x CH_4 + $(1 - x)$ SiH_4) and 5% (x CD_4 + $(1 - x)$ SiH_4) in 95% He for $x = 0$ to 1 have been analyzed by mass spectroscopy (Catherine *et al.*, 1981) and the properties of the Si_xC_y : H films deposited from the discharges have been studied and analyzed (Catherine and Turban, 1979). If either SiH_4 or CH_4 is present in excess, the discharge composition suggests that the excitation takes place primarily in the major constituent; that is, the partitioning of the energy dissipated in the discharge is not a linear function of gas composition.

Ion – molecule reactions between the two hydrides do occur since the mixed ions $Si_xC_yH_z^+$ are observed. The most abundant ions are $SiCH_3^+$, $SiCH_4^+$, and $SiCH_5^+$ and the maximum abundance for each ion occurs at a different composition, somewhere in the range 40 – 80% SiH_4 (Catherine *et al.*, 1981). The mixed ions $Si_xC_yH_z^+$ for x, $y = 1$, 2, and 3 have also been observed by sampling through the anode of a dc discharge excited in mixtures of CH_4/SiH_4 with no dilution in helium. The $SiCH_5^+$ ion is dominant, and the next most abundant ion, $SiCH_3^+$, is less abundant by a factor of 3 to 5, depending on the discharge conditions.

The ion – molecule reactions of CH_4 – SiH_4 mixtures have been studied using high-pressure tandem mass spectroscopy (Cheng *et al.*, 1973). It was found that the predominant reaction of all primary methane ions with SiH_4 was to produce SiH_3^+ by hydrogen abstraction. The production of $SiCH_5^+$ proceeds primarily via reactions of CH_3^+ and CH_4^+ with SiH_4 and the reaction of SiH_2^+ with CH_4. It was found that all primary methane ions react with monosilane, whereas the primary silane ions are much less reactive with methane. The collisionally stabilized $SiCH_7^+$ ion was not detected in the high-pressure studies.

(4) *Impurity Effects and B_2H_6 and PH_3 in SiH_4.* The persistent presence of ions having $m/e = 45, 47$, and 49 at concentrations $10^{-4} - 10^{-3}$ that of the dominant SiH_3^+ ion has been observed (Weakliem, 1982). The results of experiments involving the addition of CO_2 and H_2O as impurities and studies on the effect of substrate temperature were used to identify $m/e = 45$ as having contributions from both $SiOH^+$ and $SiCH_5^+$ and 49 as $SiH_3OH_2^+$. The identification of $m/e = 47$ is more uncertain, but we think it is due to SiF^+, which is a very stable ion and which is found to comprise at least 99% of the ion composition of a plasma excited in SiF_4. Since SiF_4 discharges were operated from time to time in the chamber and fluorine is noted for its strong adherence to stainless steel, it is reasonable to assume that the chamber walls act as a source of fluorine.

Ion–molecule reactions were also observed in rf discharges excited in $SiH_4 + B_2H_6$ and $SiH_4 + PH_3$ gas mixtures. The positive ion spectrum of a discharge in the gas mixture 1% B_2H_6 + 99% SiH_4 revealed that the concentration of $B_2H_5^+$ was almost as large as that of SiH_3^+ and the mixed ion $SiBH_4^+$ was itself approximately 1% that of SiH_3^+. The positive ion spectrum of a discharge in the mixture 4% PH_3 + 96% SiH_4 reveals a large enhancement of the phosphorus-to-silicon ratio in the ion composition compared with the neutral gas mixture. The dominant phosphorus-containing ion is PH_4^+—a stable, filled orbital sp^3, tetrahedral molecule. The enhancement of the $B_2H_y^+$ and PH_y^+ ions compared with the neutral gas composition was *not* observed in dc discharges sampled through the anode. In these cases the $(B_2H_y^+)/(SiH_n^+)$ and $(PH_y^+)/(SiH_n^+)$ ion ratios were very nearly equal to the corresponding neutral gas composition. Finally, it was found that there is a large enhancement in the incorporation of B (and P) in the solid a-Si:H films for dc deposition, but the incorporation ratio is nearly unity for rf(C)-deposited films.

7. NEUTRALS

The decomposition of silane in a glow discharge produces the higher silanes, Si_nH_{2n+2}, as well as ions and solid a-Si:H deposits. Disilane, Si_2H_6, and trisilane, Si_3H_8, are commonly observed (Turban *et al.,* 1980; Weakliem, 1981), but higher silanes with n up to 6 have been observed (Gallagher and Scott, 1982). There is a rapid decrease in the relative concentration as the size of the molecule increases. The relative concentrations are typically $SiH_4 : Si_2H_6 : Si_3H_8 = 100 : 10 : 1$.

Each molecule has a characteristic fragmentation pattern, which is given in Table I for SiH_4, Si_2H_6, and Si_3H_8. The contribution of the $Si_2H_y^+$ and SiH_y^+ fragments arising from the dissociative ionization of Si_3H_8, have not been determined, and Table I simply lists the parent $Si_3H_y^+$ contributions. The trisilane produced during the operation of a silane glow discharge was used to obtain these parent $Si_3H_y^+$ ion abundances. The values listed for Si_2H_6 were directly measured using pure disilane (Potzinger and Lampe, 1969). The disilane fragmentation pattern in the electron energy range 10–150 eV was recently measured by Gallagher and Scott (1983) and independently in my laboratory. These results are generally in good agree-

TABLE I

FRAGMENTATION PATTERNS OF SiH_4, Si_2H_6, AND Si_3H_8 FOR 70-eV ELECTRON ENERGY[a]

m/e	Ion	SiH_4	Si_2H_6[b]	Si_3H_8[c]
28	Si^+	28		
29	SiH^+	29	23	
30	SiH_2^+	100	10	
31	SiH_3^+	79	19	
56	Si_2^+		30	
57	Si_2H^+		41	
58	$Si_2H_2^+$		64	
59	$Si_2H_3^+$		21	
60	$Si_2H_4^+$		100	
61	$Si_2H_5^+$		31	
62	$Si_2H_6^+$		52	
84	Si_3^+			73
85	Si_3H^+			100
86	$Si_3H_2^+$			45
87	$Si_3H_3^+$			36
88	$Si_3H_4^+$			72
89	$Si_3H_5^+$			46
90	$Si_3H_6^+$			96
91	$Si_3H_7^+$			48
92	$Si_3H_8^+$			38

[a] Abundances normalized to 100 for most abundant ion of each parent molecule.

[b] From Potzinger and Lampe (1969).

[c] Measurements of the trisilane produced in both dc and rf glow discharges in silane.

ment with each other and with the 70-eV values given in Table I. However, our measurement for the 70-eV relative abundance of SiH^+ ($m/e = 29$) from disilane fragmentation is about one third the value given in Table I.

Kinetic studies of the decomposition of silane in a static dc silane discharge have shown that the silane decomposition rate and the disilane and trisilane generation rates are independent of pressure, and linearly dependent on discharge current, at least for low current (Longeway et al., 1983). At higher currents, when a significant fraction of the gas composition is disilane, which may itself be decomposed by the discharge, the fraction of higher silanes in the gas tends to saturate.

The principal reaction pathways for the production of disilane, trisilane, and the higher silanes involves the silyl radical, SiH_3, and the silylene radical, SiH_2, and these reactions are discussed in Chapters 8 and 9. There is a growing consensus that, except for cathodic films and very low-pressure depositions, the film formation occurs via free radicals.

The total dissociation cross sections for silane and disilane were recently determined from a kinetic analysis of a multipole discharge (Perrin et al., 1982). It was found that for electron energies greater than 50 eV the ratio of the total ionization cross sections to the total cross sections for dissociation into neutral fragments was unity for silane and $\frac{1}{3}$ for disilane.

The existence of SiH_2 and SiH_3 free radicals in the discharge was deduced from a mass spectroscopic study of the gaseous products of $SiH_4 - SiD_4 - He$ and $SiH_4 - H_2 - D_2$ discharges and from the infrared spectra of the deposited films (Turban et al., 1981). The disilane formed, $Si_2H_{6-n}D_n$, exhibited all degrees of deuteration, $n = 0 - 6$. SiH_2D_2 and $SiHD_3$ were directly detected using the known fragmentation patterns of SiH_4 and SiD_4. It was also concluded that there was not a direct correlation between SiH_n radicals in the plasma and SiH and SiH_2 vibrational groups found in the solid.

The direct detection of radicals by mass spectroscopy presents some special difficulties. Turban et al. (1980) noted that radicals could, in principle, be detected by using electron energies less than the appearance potential for an ion created by electron impact dissociation of silane but greater than the direct ionization of the same ion from the corresponding radical. For example, SiH_2^+ may be produced from SiH_4 by (III) or from SiH_2 by (XVI):

$$e + SiH_4 \rightarrow SiH_2^+ + H_2 + 2e \quad (E_a = 11.9 \text{ eV}), \quad \text{(III)}$$

$$e + SiH_2 \rightarrow SiH_2^+ + 2e \quad (I = 9.7 \text{ eV}). \quad \text{(XVI)}$$

The ionization potentials for the radicals have not been measured but may be calculated using existing thermodynamic data (Moortgat, 1970) and are given in Table II. The difference between the appearance potential and the corresponding ionization potential is $2 - 6$ eV.

Since the radicals are quite reactive and since the degree of dissociation of

TABLE II

IONIZATION POTENTIALS OF THE SiH_n RADICALS
I AND THE APPEARANCE POTENTIALS E_a OF THE
SAME ION PRODUCED BY ELECTRON IMPACT
DISSOCIATION OF SiH_4

Radical	Ion	I (eV)	E_a (eV)
SiH_3	SiH_3^+	8.2	12.3
SiH_2	SiH_2^+	9.7	11.9
SiH	SiH^+	9.6	16.1
Si^a	Si^+	8.15	11.7

[a] Direct determination from the series limit of the absorption spectrum of Si.

silane is usually not large, the direct detection of the radicals in the presence of the relatively large number of ions produced by the dissociative ionization of silane is difficult. Two aspects of this difficulty are readily apparent by examination of Table II and Fig. 5. In the first place, since E_a and I differ by only a few electron volts, the electron source should have a narrow energy spread in order to minimize overlapping contributions. In the second place, since the electron energy employed is low, the segment of the ionization cross-section curve being utilized is in the threshold region, which may be $10^{-3}-10^{-4}$ times the maximum value, and thus the ion signal is quite low.

The radicals SiH_2 and SiH_3 have been detected and measured as a function of discharge current using an indirect procedure in a 3% SiH_4 in Ar glow discharge (Bourquard et al., 1981). The quantity of silane decomposed was related to the quantity of hydrogen produced and the calculated fraction of undissociated silane was used in order to subtract its fragmentation pattern from the experimentally measured pattern. The difference was taken to be due to the radical fragments. The radical concentration was found to be $SiH_2 > SiH_3 > Si > SiH$. The procedure used depends on several assumptions whose validities are difficult to verify.

Direct mass spectroscopic detection of the radicals SiH_y, $y = 0-3$ was recently achieved using silane–argon mixtures in both rf and dc silane discharges (Robertson et al., 1983). The apparatus was designed to utilize high-flow velocity, high power and sampling close to the discharge region, in order to maximize the radical signals. The sampled species were ionized in a high-vacuum chamber by a magnetically confined, nearly monoenergetic electron source that was designed to minimize secondary electron ionization and was operated at energies near threshold. The ionized radicals were focused and mass analyzed by a quadrupole mass filter. The radical densities were found to be low; the ratio of the radical density to the reacted silane density ranged from a lower detection limit of 1×10^{-5} to a maximum of 1×10^{-4}. The Si radical was found to be dominant in high-power discharges, whereas the silyl radical, SiH_3, was found to be most abundant in discharges operated at low-power conditions typically employed for the preparation of good-quality a-Si:H films. In the latter case, the radical concentrations were found to be in the order $SiH_3 > SiH > Si > SiH_2$ and the sampling was done in the region of the substrate of a dc proximity discharge.

The difficulties encountered in radical detection are clearly described in the article by Robertson et al. (1983). Although quantitative measurements are certainly difficult, improvements may be achieved using other methods, e.g., retarding potential difference (RPD) or modulated beam mass spectroscopy, as discussed by Foner (1966). We anticipate that radical detection and detailed measurements will be made soon for a variety of discharge conditions. Such measurements, together with the measurement of various neutral molecule–silane ion reaction cross sections as a function of energy and measurements of the energy distributions of charged particles arriving at surfaces, are crucial to a good understanding of the a-Si:H film-forming process.

ACKNOWLEDGMENTS

I am grateful to Alan Gallagher and J. Scott (JILA, Boulder, CO) and to J. Perrin, J. P. M. Schmitt, G. de Rosny, B. Drevillon, J. Huc, and A. Lloret (Ecole Polytechnique, Palaiseau, France) for permission to use the figures cited.

REFERENCES

Allen, W. N., Cheng, T. M. H., and Lampe, F. W. (1977). *J. Phys. Chem.* **66**, 3371.

Austin, W. E., Holme, A. E., and Leck, J. H. (1976). *In* "Quadrupole Mass Spectrometry" (P. H. Dawson, ed.), Chapter 6. Elsevier, Amsterdam.

Bourquard, S., Erni, D., and Mayor, J. M. (1981). *In* "Proceedings of the Fifth International Symposium on Plasma Chemistry" (B. Waldie and J. A. Farnell, eds.), p. 664. Heriot-Watt University, Edinburgh, Scotland.

Carlson, D. E. (1977). U. S. Patent 4,064,521.

Catherine, Y., and Turban, G. (1979). *Thin Solid Films* **60**, 193.

Catherine, Y., Turban, G., and Grolleau, B. (1981). *Thin Solid Films* **76**, 23.

Chapman, B. (1980). "Glow Discharge Processes." Wiley, New York.

Chatham, H., Hils, D., Robertson, R., and Gallagher, A. (1983). *J. Chem. Phys.* **79**, 1301.

Chen, F. F. (1965). *In* "Plasma Diagnostic Techniques" (R. H. Huddlestone and S. L. Leonard, eds.), Chapter 4. Academic Press, New York.

Cheng, T. M. H., Yu, T.-Y., and Lampe, F. W. (1973). *J. Phys. Chem* **77**, 2587.

Coburn, J. W., and Kay, E. (1972). *J. Appl. Phys.* **43**, 4965.

Davis, W. D., and Vanderslice, T. A. (1963). *Phys. Rev.* **131**, 219.

Drevillon, B., Huc, J., Lloret, A., Perrin, J., de Rosny, G., and Schmitt, J. P. M. (1980). *Appl. Phys. Lett.* **37**, 646.

Drevillon, B., Huc, J., Lloret, A., Perrin, J., de Rosny, G., and Schmitt, J. P. M. (1981). *In* "Proceedings of the Fifth International Symposium on Plasma Chemistry" (B. Waldie and J. A. Farnell, eds.), p. 634. Heriot-Watt University, Edinburgh, Scotland.

Foner, S. N. (1966). *Adv. At. Mol. Phys.* **2**, 385.

Francis, G. (1960). "Ionization Phenomena in Gases." Academic Press, New York.

Gallagher, A., and Scott, J. (1981). *SERI Final Contract Rep.* **XJ-0-9053-1**.

Gallagher, A., and Scott, J. (1982). *SERI Final Contract Rep.* **DB-2-02189-1**.

Gallagher, A., and Scott, J. (1983). *SERI Final Contract Rep.* **XB-2-02085-1**.

Goodyear, C. C., and von Engel, A. (1962). *Proc. Phys. Soc., London* **79**, 732.

Haller, I. (1980). *Appl. Phys. Lett.* **37**, 282.

Hamasaki, T., Kurata, H., Hirose, M., and Osaka, Y. (1980). *Appl. Phys. Lett.* **37**, 1084.

Hanak, J. J., and Korsun, V. (1978). *Conf. Rec. IEEE Photovoltaic Spec. Conf.* **13**, 780.

Hasted, J. B. (1975). *Int. J. Mass Spectrom. Ion Phys.* **16**, 3.

Henis, J. M. S., Stewart, G. W., Tripodi, M. K., and Gaspar, P. P. (1972). *J. Chem. Phys.* **57**, 389.

Hotta, S., Tawada, Y., Okamoto, H., and Hamakawa, Y. (1981). *J. Phys. (Orsay, Fr.)* **42**, C4-631.

Knights, J. C. (1976). *Philos. Mag.* [8] **34**, 663.

Knights, J. C., Lucovsky, G., and Nemenich, R. J. (1979). *J. Non-Cryst. Solids* **32**, 393.

Kocian, P. (1980). *J. Non-Cryst. Solids* **35**, 195.

Kocian, P., Mayor, J. M., and Bourquard, S. (1979). *In* "Proceeding of the Fourth International Symposium on Plasma Chemistry" (S. Veprek and J. Hertz, eds.), p. 663. Univ. of Zurich, Zurich, Switzerland.

Longeway, P. A., Estes, R. D., and Weakliem, H. A. (1983). *J. Phys. Chem.* (to be published).

Matsuda, A., and Tanaka, K. (1982). *Thin Solid Films* **92**, 171.

Moortgat, G. K. (1970). Ph.D. Thesis, University of Detroit, Detroit, Michigan.

Mosburg, E. R., Kerns, R., and Abelson, J. (1982). *In* "Plasma Processing, Third Symposium" (J. Dielman, R. G. Frieser, and G. S. Mathad, eds.), p. 410. Electrochem. Soc., Pennington, New Jersey.

Mosburg, E. R., Kerns, R. C., and Abelson, J. R. (1983). *J. Appl. Phys.* (to be published).

Mott, N. F., and Massey, H. S. W. (1965). "Theory of Atomic Collisions," 3rd ed. Oxford Univ. Press, London and New York.

Okamoto, H., Nitta, Y., Adachi, T., and Hamakawa, Y. (1979). *Surf. Sci.* **86**, 486.

Perrin, J., and Schmitt, J. P. M. (1982). *Chem. Phys.* **67**, 167.

Perrin, J., Schmitt, J. P. M., de Rosny, G., Drevillon, B., Huc, J., and Lloret, A. (1982). *Chem. Phys.* **73**, 383.

Potzinger, P., and Lampe, F. W. (1969). *J. Phys. Chem.* **73**, 3912.

Purdes, A. J., Bolker, B. F. T., Bucci, J. D., and Tisone, T. C. (1977). *J. Vac. Sci. Technol.* **14**, 98.

Robertson, R., Hils, D., Chatham, H., and Gallagher, A. (1983). *Appl. Phys. Lett.* (to be published).

Taniguchi, M., Hirose, M., Hamasaki, T., and Osaka, Y. (1980). *Appl. Phys. Lett.* **37**, 787.

Turban, G., Catherine, Y., and Grolleau, B. (1979). *Thin Solid Films* **60**, 147.

Turban, G., Catherine, Y., and Grolleau, B. (1980). *Thin Solid Films* **67**, 309.

Turban, G., Catherine, Y., and Grolleau, B. (1981). *Thin Solid Films* **77**, 287.

Turban, G., Catherine, Y., and Grolleau, B. (1982). *Plasma Chem. Plasma Process.* **2**, 61.

Vasile, M. J., and Smolinsky, G. (1975). *Int. J. Mass Spectrom. Ion Phys.* **18**, 179.

Veprek, S. (1976). *Pure Appl. Chem.* **48**, 163.

Veprek, S., and Marecek, V. (1968). *Solid-State Electron.* **11**, 683.

Vossen, J. L., and Cuomo, J. J. (1978). *In* "Thin Film Processes" (J. L. Vossen and W. Kern, eds.), Chapter II-1. Academic Press, New York.

Waits, R. K. (1978). *In* "Thin Film Processes" (J. L. Vossen and W. Kern, eds.), Chapter 2, p. 4. Academic Press, New York.

Weakliem, H. A. (1980). *AIAA Pap.* **80-1327.**

Weakliem, H. A. (1982). *In* "Plasma Processing, Third Symposium" (J. Dielman, R. G. Frieser, and G. S. Mathad, eds.), p. 14. Electrochem. Soc., Pennington, New Jersey.

Yu, T.-Y., Cheng, T. M. H., Kempter, V., and Lampe, F. W. (1972). *J. Phys. Chem.* **76,** 3321.

Structure

SEMICONDUCTORS AND SEMIMETALS, VOL. 21A

CHAPTER 11

Relation between the Atomic and the Electronic Structures[†]

Lester Guttman

MATERIALS SCIENCE AND TECHNOLOGY DIVISION
ARGONNE NATIONAL LABORATORY
ARGONNE, ILLINOIS

I. Introduction

From well-established principles it is known that the electronic structure of a solid is determined by the locations of the atoms that make it up. The purpose of this chapter is to summarize the present knowledge of the atomic structure of amorphous hydrogenated silicon, a-Si:H, and to review the quantum-mechanical calculations of its electronic structure that have been based on that knowledge. The summary will be confined to the bulk structure and will therefore deal with defects only incidentally; the latter are the subject of Chapter 14 by Adler, this volume.

In contrast to the case of crystals, the structure of a substance like a-Si:H

† Work supported by the U.S. Department of Energy.

must necessarily be very incompletely known. For a perfect crystal, the relative positions of any number of atoms are determined, given a few parameters, namely, the size and shape of the unit cell and the locations of the atoms within the cell. The absence of long-range order in a-Si:H prevents such a concise description, and we must be content to know only the arrangements of atoms that are rather close to each other. But even this information is available only in probabilistic terms (i.e., as distribution functions), because the disorder of an amorphous phase consists in part of spatial fluctuations in interatomic distances. Moreover, only pair distribution functions can presently be measured with any accuracy. Those connecting three or more atoms, which would contain information about angular distributions, must be deduced with the aid of auxiliary assumptions whose validity may be questionable.

II. The Atomic Structure of Amorphous Hydrogenated Silicon

1. Diffraction by Amorphous Solids

The elastic scattering of waves (x rays, electrons, or neutrons) from spherical atoms in a noncrystalline solid can be treated in an elementary way as follows (for a complete treatment, see Wagner, 1979). The *amplitude* of the scattered wave at some distance from the sample is

$$A(\mathbf{k}) = \sum_{n=1}^{N} b_n \exp(i\mathbf{k} \cdot \mathbf{r}_n). \tag{1}$$

Here $\mathbf{k}$, the scattering vector, is the difference between the incident wavevector and the scattered wave vector, and its magnitude is $k = 4\pi \sin \theta/\lambda$, where θ is the Bragg angle. The wavelengths of the incident and the scattered radiation λ are the same since the scattering is elastic. The scattering is also assumed to be coherent (i.e., phase preserving), so that the amplitudes of the waves scattered by the individual atoms simply add, as in Eq. (1). The coefficients b_n depend on the nature of the scatterer and on the type of radiation. For x rays and electrons, b is the atomic scattering factor, which is k dependent. For neutrons, b is the coherent scattering length and is independent of k. Assume first that there are N identical atoms in a sample of volume V. Then the scattered intensity per atom is proportional to the time average of $|A|^2$, that is,

$$\frac{I(\mathbf{k})}{N} = \frac{b^2}{N} \left\langle \left| \sum_{i=1}^{N} e^{i\mathbf{k} \cdot \mathbf{r}_i} \right|^2 \right\rangle_{AV}$$

and

$$\frac{I(\mathbf{k})}{Nb^2} = \frac{1}{N} \left\langle \sum_{i=1}^{N} \sum_{j=1}^{N} e^{i\mathbf{k} \cdot (\mathbf{r}_i - \mathbf{r}_j)} \right\rangle_{AV}. \tag{2}$$

The right-hand side of Eq. (2) is called the "structure factor," $S(\mathbf{k})$:

$$S(\mathbf{k}) = \frac{1}{N} \left\langle \sum_{i=1}^{N} \sum_{j=1}^{N} e^{i\mathbf{k}\cdot(\mathbf{r}_i - \mathbf{r}_j)} \right\rangle_{AV}.$$

Take out the N terms with $i = j$ in the sum, each of which is just unity:

$$S(\mathbf{k}) = \frac{1}{N} \left[N + \left\langle \sum_{i=1}^{N} \sum_{j=1\neq i}^{N} e^{i\mathbf{k}\cdot\mathbf{r}_{ij}} \right\rangle_{AV} \right].$$

Since all the atoms are equivalent, we can assume that the first summation, which is a spatial average, gives N identical contributions and that the second sum can be replaced by an integral:

$$S(\mathbf{k}) = 1 + \int_V e^{i\mathbf{k}\cdot\mathbf{r}} n(\mathbf{r}) \, d^3\mathbf{r},$$

where the origin is excluded from the integration. Here $n(\mathbf{r})$ is the average number of atoms per unit volume at a point $\mathbf{r}$, measured from any atom. In an isotropic glass, n depends only on the magnitude of $\mathbf{r}$, not on its orientation, and for large interatomic separations approaches its random value, the average number density ρ_0,

$$\rho_0 = N/V$$

Defining a pair-distribution function $g(r)$ as the ratio of $n(r)$ to ρ_0, we get

$$S(k) = 1 + \rho_0 \int_V e^{i\mathbf{k}\cdot\mathbf{r}} g(r) \, d^3\mathbf{r}. \tag{3}$$

Adding and subtracting 1 in (3), we get

$$S(k) = 1 + \rho_0 \left\{ \int_V e^{i\mathbf{k}\cdot\mathbf{r}} [g(r) - 1] \, d^3\mathbf{r} + \int_V e^{i\mathbf{k}\cdot\mathbf{r}} \, d^3\mathbf{r} \right\}.$$

The second integral vanishes unless $\mathbf{k} = 0$, when it gives a constant N, which we may drop, since it is an unobservable scattering in the direction of the transmitted beam.

Since $g(r)$ goes to 1 as $r \to \infty$, the remaining integral is well behaved, and the relationship can be inverted to give the pair-distribution function as the Fourier transform of the observed scattering. This is the only distribution function for which a rigorous experimental determination is possible. We see from the derivation that this function is singled out because the scattering is determined by the interference of waves scattered from *pairs* of atoms.

The case of most interest to us is that of a *binary* glass. Then there are two distinct coherent scattering lengths, b_1 and b_2, for the two kinds of atoms, and three pair-distribution functions, $g_{11}(r)$, $g_{22}(r)$, and $g_{12}(r)$, for the probabilities of finding two atoms at a distance r from each other, both of type 1, both of type 2, or one of each type, respectively. Correspondingly, there exist three structure factors, $S_{11}(k)$, $S_{12}(k)$, and $S_{22}(k)$. Analogously to Eq. (2), we can write for the coherent scattering

$$I(k)/N = x_1 b_1^2 S_{11}(k) + 2\sqrt{x_1 x_2}\, b_1 b_2 S_{12}(k) + x_2 b_2^2 S_{22}(k). \tag{4}$$

The atomic fractions of the two species are x_1 and x_2. For a-Si:H the subscripts 1 and 2 refer to silicon and hydrogen, respectively. However, now the three pair-distribution functions cannot be calculated by a simple Fourier transformation of (4). Even if the scattering lengths are k independent, we obtain from an experimental measurement of $I(k)$ only a linear combination of the three S_{ij}, which cannot generally be resolved into its individual components. Because of this difficulty we have very little direct information about the structure of a-Si:H itself. Instead, conjectures about its structure have been based on that of pure a-Si modified in ways that are consistent with general principles and with various experimental evidence, mainly spectroscopic. We turn next to the question of the structure of pure a-Si.

2. EXPERIMENTAL DETERMINATION OF THE ATOMIC STRUCTURE OF AMORPHOUS SILICON

The preparation of high-quality a-Si in the quantities that are necessary for x-ray diffraction experiments is not easy, and even larger amounts are needed for neutron diffraction. Since amorphous germanium is more readily available, more work has been done on it, and its structure is precisely known. Crystalline silicon and germanium resemble each other closely in all their properties, so it has generally been assumed that their structures in the amorphous form are also similar, and this is borne out by the few existing studies of the former. It is convenient, therefore, to discuss the structure of a-Si as if it were the same as that of a-Ge, with the reservation that there may be differences in detail.

Figure 1 shows the result of a recent very thorough study of a-Ge by neutron diffraction (Etherington *et al.*, 1982). This result is typical of diffraction studies of a-Ge or a-Si and will serve to illustrate the data on which are based current ideas about their (common) structure. The ordinate in Fig. 1 is the experimental function $t'(r)$, which is the convolution of the total correlation function

$$t(r) = 4\pi r \rho_0 g(r) \tag{5}$$

with a "peak function" that expresses the effects of the finite range of the data. The solid line is this function after the correction for sample contamination by about 0.3 mol % of water. The first peak is centered at 2.463 Å, and the area under this peak is 3.68. These numbers are generally interpreted to mean that nearly every Ge atom has the same number (4) of

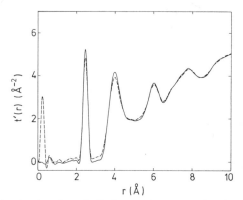

FIG. 1. Pair-distribution function of a-Ge from neutron scattering. (From Etherington *et al.*, 1982.)

neighbors at almost exactly the same distance (2.450 Å) as it has in the crystal. The second peak in $t'(r)$, at about 4 Å, is not completely resolved from the next, but if it is assumed that only second neighbors contribute on the small r side, there would be 12.1 of these, instead of 12 exactly, at an average distance only 0.025 Å less than that in the crystal. The width of the first peak is primarily instrumental, because the data terminate at $k_{max} \cong 24 \text{ Å}^{-1}$. The remaining width can be entirely accounted for by thermal broadening, that is, the static variation in first-neighbor distance is too small to be measured by neutron diffraction (also see Rabe *et al.*, 1979). The width and position of the second peak are consistent with an rms variation of 9.7° in interbond angle about an average value of 108.5 ± 1°, instead of $\cos^{-1}(-\frac{1}{3}) = 109.5°$ as in the crystal.

The locations and numbers of atoms in the first two coordination shells of a germanium atoms as deduced from this work are in very good agreement with earlier x-ray and electron diffraction measurements on a-Ge, within the somewhat lower precision of the latter. With only a change of scale, they hold also for the experimental studies of amorphous silicon (Moss and Graczyk, 1970). Beyond the first two shells, however, it is not possible to associate maxima in $g(r)$ with clearly separable coordination shells. The data obviously are not consistent with a random distribution of atoms, but the structural information buried in the features of $g(r)$ at intermediate distances can only be extracted in conjunction with a *model*. In the next section, we discuss existing models and their validity.

3. MODELS OF THE ATOMIC STRUCTURE OF AMORPHOUS SILICON

In the preceding section the atomic structures of a-Ge and a-Si were compared with those of their crystalline forms insofar as this was possible at

short distances. It has generally been accepted that a good first approximation to the bulk structures is that (a) each Ge or Si atom has exactly four
others around it at the same distance as in the crystal, forming a regular
tetrahedron *on the average,* and (b) each of these in turn has three more
neighbors, similarly arranged, giving rise to a well-defined second shell of 12
atoms, none of which is a neighbor of any other. Models differ first on
whether or not this local order is propagated far enough to produce recogniz-
able microcrystals. In my opinion, microcrystalline models suffer from so
many difficulties that they can be rather summarily dismissed. To survive at
all, it is necessary for them to account for the large fluctuations in second-
neighbor distance and for the absence of Bragg reflections in purely ad hoc
ways, and in no case has it been shown that microcrystalline models have
any advantages that justify the effort needed to make them fit the facts.

The alternate view, that the resemblance of the amorphous form to the
crystal ceases beyond the second shell, is overwhelmingly more popular and
has been realized in a number of *random-network* models. The random-network concept, as proposed first by Zachariasen (1932), was applied by him
to certain oxide glasses for precisely the reason that their diffraction patterns
required close resemblances of their structures to those of the corresponding
crystalline forms at the local level. He therefore suggested that the numbers
and kinds of first neighbors were the same in both forms but that variations
in the interbond angles among these neighbors led rapidly to a loss of local
order and ultimately to the absence of long-range order (i.e., to the absence
of crystallinity).

The first step in the construction of any random-network model of a-Si is
to establish its *topology.* That is, having adopted the view that every atom is
to have perfect four-fold coordination, one must identify those four atoms
that are the first neighbors of each atom. In the first models of this kind that
were built, an atom (represented by a ball) was connected to four others by a
mechanism that permitted very little variation in interatomic distance but
did allow about 10° variation in interbond angle. Atoms were added singly
wherever those present were not yet fully coordinated and were positioned
so as to minimize the angular distortions without falling into the trap of
crystallinity. When the model was large enough, its *geometry* was determined by direct measurement. It was soon realized that although a macroscopic model made in this way might satisfactorily represent the bond
pattern of the atomic system and might even approximate its scaled atomic
positions, the mechanical properties of the model could not be translated
into atomic terms. Such a model could therefore only serve as a starting
point for quantitative studies of real a-Si.

The equilibrium atomic positions in a-Si are those that minimize its
potential energy, which is, by the Hellman–Feynman theorem, the total

energy of the electrons in the field of the Si nuclei plus the internuclear repulsive energy. This potential is, of course, altogether unknown and of a very complex form for general atomic positions. However, for atomic configurations not too far from those of the static structure, it may be warranted to assume, as has been done by everyone in the field, that the potential function has a simple valence-force form. That is, the potential energy is assumed to be a function only of the deviations of the first neighbor distances from the normal bond length and of the deviations of the inter-bond angles from the tetrahedral angle. Without exception this function has been taken to be the Keating (1966a,b) potential,

$$V = \frac{1}{2} \sum_{l=1}^{N} \left[\frac{3\alpha}{4d^2} \sum_{i=1}^{4} (r_{li}^2 - d^2)^2 + \frac{3\beta}{2d^2} \sum_{i>j=1}^{4} \left(\mathbf{r}_{li} \cdot \mathbf{r}_{lj} + \frac{d^2}{3} \right)^2 \right]. \qquad (6)$$

Here $\mathbf{r}_{li}$ denotes the vector from atom l to its first neighbor atom i, d is the equilibrium bond length, and α and β are two interatomic force constants. This function gives the correct values (within 2%) for the three elastic moduli of crystalline C, Si, and Ge with the choice of only two parameters, α and β. It is convenient that β/α has nearly the same value (0.3) for both Si and Ge, so that except for a change of distance scale, a model can represent a-Ge or a-Si equally, if the equilibrium positions have been derived from the Keating potential. It should also be remarked (a) that although Eq. (6) appears to be a quartic function, it actually reduces to a quadratic function (i.e., harmonic) for small deviations of the distances and angles from their ideal values and (b) that the validity of (6) for the elastic range of crystalline Si and Ge is no guarantee that it holds for the much larger distortions found in the amorphous state. A more general form devised by Keating also reproduces the third-order elastic constants of the crystals and should therefore be better for the amorphous forms, but it has never been applied there. At this point, it is also worth noting that irrespective of the particular form chosen, any potential function of the valence – force type has the effect of making the equilibrium *geometry* of a model dependent only on its *topology:* Specifying which atoms are bonded to each determines that certain interatomic distances and interbond angles appear in the potential function and no others.

The first random network model of the Group IV elements was built by Polk from mechanical units (Polk, 1971; Polk and Boudreaux, 1973). It was later enlarged from 440 to 519 atoms, and the atomic coordinates were adjusted by computer to minimize the variation in first-neighbor distance. Subsequently, the latter model and another containing 201 atoms were simulated by computer and relaxed under the Keating potential (Steinhardt *et al.,* 1974). A mechanical model was constructed by Connell and Temkin

(1974) that, unlike the preceding examples, contained no odd-numbered rings of bonds, and could therefore also represent the amorphous III–V semiconductors without forcing first neighbors to be of the same species. Another approach to construction of a mechanical model is that of Evans *et al.* (1974), who noted that a-Si would result formally if the oxygen atoms were removed from a model of a-SiO$_2$, and the remaining Si atoms were connected. The flexibility of the Si–O–Si bridges in a-SiO$_2$ permits the formation of rings containing as few as three or four Si atoms and three or four O atoms; if these rings are present in a-SiO$_2$, as is likely, removal of oxygen from a good model of that material would lead to a poor model of a-Si, which would contain numerous three- or four-membered, highly strained rings of Si. (The converse is true of attempts to model a-SiO$_2$ simply by putting oxygen atoms between pairs of Si atoms in a realistic model of a-Si.)

The limitations of mechanical models for quantitative studies are evident, even when they are supplemented by computer simulation. The construction process is tedious and fraught with subjective elements, and although the final product may contain several hundred atoms, many are near or at a free surface. It was soon realized that these faults could be alleviated if the construction process itself could be computerized. The first step was taken by Henderson and Herman (1972) and Henderson (1974), who constructed single examples of computer models that satisfied periodic boundary conditions.† Because these models had no free surfaces, their small sizes (64 or 61 atoms in the repeating unit) were expected to be a less serious failing than in nonperiodic models. Automatic methods of generating finite models have been described by Duffy *et al.* (1974), who used them to construct a variety of examples. A fully automatic process capable in principle of generating periodic models of any size was devised by Guttman (1974, 1976). This method can generate models with angular distortions about 10% greater than observed and with fluctuations in first-neighbor distance of about 2–3%. A partly automatic procedure has been used by Beeman and Bobbs (1975) to modify the topology of a finite model so as to bring its pair distribution function closer to that observed. Sadoc and Mosseri (1982) have proposed construction of models whose topology is determined by a "tiling" of four-dimensional polytopes; no comparison with experiment has yet been published of the resulting pair-distribution functions, and it is not clear what physical principle underlies this exotic scheme.

---

† The use of periodic boundary conditions is standard in computer simulations of phenomena on the atomic scale. The "system" is taken to be contained in a rectangular parallelepiped ("box") that is surrounded by copies of itself on all sides. The "box" is defined by its dimensions, but its location is arbitrary, and all of the atoms in the system are effectively "interior" atoms.

The models produced by the foregoing methods have been compared to their experimental results for a-Ge by Etherington *et al.* (1982), and we cannot do better than to summarize their main conclusions. The excellent original paper should be consulted for details. Briefly, these authors concur in rejecting quasi-crystalline models (see also Weinstein and Davis, 1973), as well as that derived from a-SiO$_2$, as showing too much structure. The Henderson model, however, shows too little. The remaining (finite) random-network models agree well with experiment, but all tend to show too much structure at large distances. We will henceforth adopt the viewpoint that the random-network model is an adequate representation of a-Si. If discrepancies exist for a-Si, they are about at the level of the differences between existing models and cannot be settled until a-Si has been characterized by diffraction as well as has a-Ge.

4. EXPERIMENTAL DETERMINATION OF THE STRUCTURE OF AMORPHOUS HYDROGENATED SILICON

In taking up the subject of experimental determination of the structure of amorphous hydrogenated silicon, which is central to the remainder of this review, we are at once confronted with great complexities, because the properties of a-Si:H vary so much with the conditions of preparation. (See Chenevas-Paule, Chapter 12, this volume.) Much of this variation arises from defects such as surfaces or voids whose size is much larger than atomic dimensions and which are dealt with elsewhere. Some, however, is due to the variety of ways in which hydrogen can be incorporated into a random network, which could in principle be clarified by knowing the Si–H pair distribution function. As noted in Section 1, that function cannot be extracted from any single diffraction measurement, the results of which depend also on the two distribution functions for the pairs of the other two types, Si–Si and H–H. There has been, in fact, one attempt to determine these functions by neutron diffraction (Postol *et al.,* 1980). It made use of the fortunate circumstance of the availability of two isotopes of hydrogen with quite different neutron scattering lengths. Measurements of the diffracted intensity from large sputtered samples of pure a-Si and of a-Si:H and a-Si:D having equal concentrations of hydrogen and deuterium, respectively, could be analyzed to yield fairly accurate scattering functions $S_{Si-Si}(k)$ and $S_{Si-H(or D)}(k)$.

Most of the current view of the structure of a-Si:H derives from its infrared absorption spectrum. Depending on the temperature of deposition, the composition of the ambient gas, and many other parameters (see the article by Zanzucchi in Chapter 4 of Volume 21B, this treatise) the amorphous film may contain up to 40–50 at. % of hydrogen. The vibrational spectra of these films, when compared with those of gaseous silanes, seem to show that much of the hydrogen is attached to silicon by ordinary covalent bonds and that a silicon atom may be bound to one, two, or three hydrogen atoms. These atomic groups are referred to as mono-, di-, or trihydride, respectively. A dissent from this interpretation has been expressed by Paul (1980), but it is otherwise generally accepted and will be adopted here.

There is also experimental evidence that, depending on how it is made, a-Si:H may have regions of diameter 10–100 Å, some of which are poorer and some richer in hydrogen than the nominal composition (Reimer et al., 1981; Carlos and Taylor, 1980; Knights and Lujan, 1979; D'Antonio and Konnert, 1979; Postol et al., 1980). For the purpose of treating the electronic structure of a-Si:H at the level of this chapter, this complication must be disregarded.

5. Models of the Atomic Structure of Amorphous Hydrogenated Silicon

The structural principles indicated by the results of the preceding section have actually been used to construct models in only two instances. A mechanical model containing 314 Si atoms and 83 H atoms was made by Weaire et al. (1979). Weaire and Wooten (1980) relaxed this model with a modified Keating potential and found agreement between the Si–Si pair-distribution function and that calculated for pure a-Si by Steinhardt et al. (1974). The rigidity of the mechanical units was such that angular distortions greater than 10° were prevented. This had the effect that the presence of a single hydrogen atom required additional hydrogen atoms on Si in its vicinity in order to satisfy the fourfold coordination of Si. The hydrogen atoms were thus clustered, and their pair-distribution function showed structure. The Si–H pair function was not computed.

In the other instance (Guttman, 1981), 12 examples of a-Si:H as monohydride were constructed by attaching hydrogen atoms to broken bonds in existing periodic models of pure a-Si. A stereoscopic view of a portion of one of these models is shown in Fig. 2. The only rule followed was that H atoms should not be attached to Si atoms that were first neighbors; the choice of bonds to be broken was otherwise random, although an attempt was made to choose those that would most effectively lower the average angular distortion. The H–H pair function therefore showed no structure (i.e., no clustering was evident). The Si–Si structure factor was in good agreement with the results of Postol et al. (1980). The Si–H structure factor also fitted the principal features of that observed by them, but differences were detectable, indicating the need for refinement of the model. (See Fig. 3.†)

† There is also a need for further neutron diffraction studies of a-Si and a-Si:H to extend the data to larger values of k, and, if possible, to improve the data at small k, which is obscured by scattering from gross defects in the sputtered samples of Postol et al.

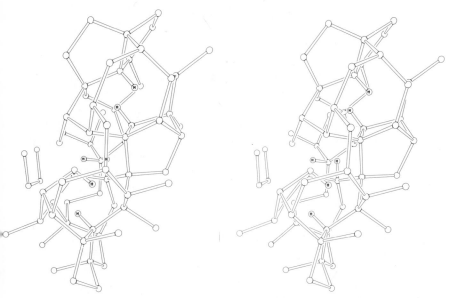

FIG. 2. Stereoscopic views of a portion of a periodic model of monohydride a-Si:H. Only the region near the hydrogen atoms is shown. (From Guttman, 1981.)

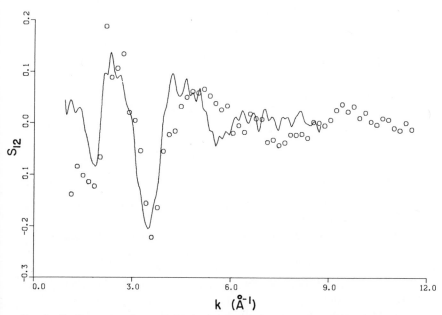

FIG. 3. Si–H structure factor. Solid line, observed (Postol *et al.,* 1980); points, calculated (Guttman, 1981.)

III. The Electronic Structure of Amorphous
Hydrogenated Silicon

6. THE TASK OF THEORY

From any viewpoint a reliable calculation of the electronic structure of a-Si: H is highly desirable. As a prototypical amorphous material, its properties differ from those of crystalline silicon in ways that we would like to understand basically. As a cheap semiconductor, its usefulness is likely to be fully exploited only if intrinsic and extrinsic properties can be separated, and the history of crystalline semiconductors is replete with instances of how theory can help to do this. However, we cannot expect that quantum theory will produce as detailed an electronic structure as in crystalline semiconductors, because the absence of long-range order in a-Si makes the Brillouin zone scheme meaningless. Rather, we can only expect to calculate average properties, foremost among them being the total density of one-electron states as a function of electronic energy. Further, one might hope to relate the differences in electronic structure of the amorphous and crystalline forms to specific differences in their atomic structures and possibly to compute important properties such as the optical absorption spectrum and transport coefficients.

7. THEORETICAL RESULTS FOR PURE AMORPHOUS SILICON

Since the random-network concept in conjunction with a simple interatomic potential function provides a moderately satisfactory description of the atomic structure of a-Si, theories of its electronic structure have generally been based on this model. However, no expression of the model in analytical form exists, nor would there be any way to carry out a calculation of the electronic structure based on such an analytical expression if it did exist. Therefore actual calculations have retained only such parts of the model as are tractable, leaving uncertain the effects of what has been neglected when it comes to comparison with the real material.

a. Topological Models

In one approach only the *topological* aspect of the model has been retained. That is, a-Si has been treated as if it differed from crystalline Si only in the presence of five-, seven-, eight-membered rings, etc., of nearest-neighbor bonds, in addition to the six-membered rings that exist exclusively in the crystal. These treatments, which stress "topological disorder," can treat infinite systems, but only at the expense of ignoring the concomitant "positional disorder," that is, the deviations of interatomic distances from those in the crystal. The results obtained by use of this simplification have been adequately summarized by Thorpe and Weaire (1974) and later by

Mott and Davis (1979), so only a few generalities need to be mentioned here. First, the density of states for a fully four-coordinated network can be expected to be nonzero in two energy regions, a filled "valence band" and an empty "conduction band" separated by a region of zero density (a "gap"), even though crystalline order has been lost. Second, the density of states in the valence band shows broad features that depend on the proportions of rings of various sizes.

b. Cluster Models

Another reduction of the full random-network model that makes numerical calculations possible is to take only a finite number of atoms for consideration. Within this cluster the effects of both topological and positional disorder can be treated with an accuracy that depends mainly on the electronic Hamiltonian assumed. Necessarily, Si atoms on the surface of the cluster are bound to fewer than four other Si atoms, and unless the cluster is immense (10^3–10^4 atoms, say), the "dangling bonds" would have an intolerably large effect on the calculated electronic states. The effect can be reduced in various ways, for example, by attaching hydrogen atoms to the unpaired Si valence electrons or, better, by extending the cluster indefinitely via a medium that can be treated accurately and whose influence can be estimated. A technique of the second kind is the "Bethe lattice" method,† extensively used by Joannopoulos and co-workers, whose results are reviewed by Joannopoulos and Cohen (1976). This review should also be consulted for authoritative and detailed coverage of other work, including that mentioned in the preceding section, 7a. Joannopoulos and Yndurain (1974) calculated the density of states at several atoms selected from finite random-network models and embedded in a Bethe lattice. The Hamiltonian did not include the effects of positional disorder, and so, as might be expected, they found also the splitting into occupied and unoccupied bands separated by a gap, as well as the effects of the ring structure noted previously. Judging by comparison of the results of this method with those of band-theoretic calculations with the same Hamiltonian for various crystalline polymorphs of Si and Ge, the accuracy appears to be quite high. However, the relative heights of the principal maxima in the valence band are reversed from the experimental values (Spicer and Donovan, 1971) and from the values found by others theoretically, possibly because of the simplicity of the Hamiltonian. Finite clusters selected from the models of

---

† In this method, each dangling bond of a surface atom is joined to another Si atom, which in turn is joined to three more, and so on, in an infinitely branching pattern. This "lattice" contains no rings, since atoms are simply added in sequence, and are never joined to more than one atom already present.

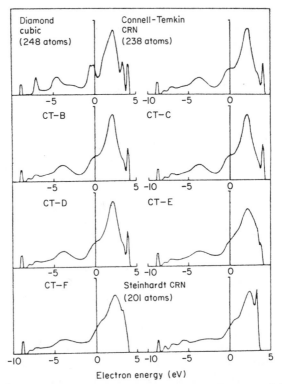

FIG. 4. Total densities of states in valence band calculated by Meek (1977) for crystalline Si, and a number of finite continuous random-network (CRN) models, including (CT–B to CT–F), the modifications of the Connell–Temkin model generated by Beeman and Bobbs (1975) containing progressively more odd-membered rings.

Polk–Boudreaux (PB) and Connell–Temkin (CT) have also been treated by Kelly and Bullett (1976) by a rather different method. The effects of the finiteness of the cluster have been tested by varying the boundary conditions. The Hamiltonian was of the "tight-binding" type; it included an interaction between electrons in atomic orbitals directed toward a pair of first neighbors that was linearly dependent on the angle between them and included as well an interaction between electrons in orbitals separated by an intervening bond that was linearly dependent on the cosine of the dihedral angle between them.† Only the valence band could be analyzed, but the

---

† It is not clear why this dependence was assumed. For first neighbors the energy should be a minimum at the tetrahedral angle, and so should vary *quadratically* near that angle. Similarly, the variation with dihedral angle θ should reflect the three fold symmetry of rotation around the intervening bond and depend on $\cos 3\theta$ rather than on $\cos \theta$.

method is capable of handling positional as well as topological disorder. The density of states again showed only broad maxima and minima instead of the sharp peaks and valleys exhibited by crystalline semiconductors. The exclusively even-membered ring CT model was found to differ in a number of secondary features from the PB model, showing once more the importance of the topology. Later calculations by Meek (1977), applying the same method to a range of models (Beeman and Bobbs, 1975) intermediate in topology between the extremes of PB and CT, confirm and extend these conclusions about the effects of topology. Figure 4, taken from Meek's paper, serves to illustrate the shape of the valence band density of states found generally in this kind of calculation, and the variation resulting as the structure is progressively modified from all even-membered rings to a mixture of even- and odd-membered rings.

c. Periodic Models

It is clear from the foregoing sections 7a and 7b, and is also evident in the original works cited, that these attempts to compute the electronic structure of a-Si are all deficient in some way, the most important being the simplicity of the Hamiltonians used, and the uncertainties about the effects of the finiteness of the underlying random-network models. As noted earlier, the second defect can be alleviated by the use of periodic models. This was first done by Ching, Lin, and Huber (1976) on the basis of the Henderson (1974) model, and later by Ching, Lin, and Guttman (1977) on the basis of a model denoted by them as G54. The Hamiltonian was such that positional disorder could be fully taken into account at a certain level of approximation, and a technique was introduced that enabled a reliable assessment to be made, incidentally, of the errors attributable to dealing with finite clusters. In all models except that of Henderson, but including those of PB and CT, a well-defined gap appeared, whose size seemed to be (negatively) correlated with the amount of positional disorder, as measured by the mean deviations of first-neighbor distance and interbond angle from their ideal values. The effects of topological disorder, on the other hand, were much less evident than in the work described in the previous sections.

In a subsequent paper, Ching and Lin (1978) calculated the optical absorption spectrum of the periodic model G54 (in good agreement with experiment) and introduced the concept of a "localization parameter" based on a population analysis of the wave functions of the model. More specifically, these wave functions, being linear combinations of 3s- and 3p-like atomic orbitals, assign to each atom a certain fraction of the charge in each occupied electronic state. If these charges were equal, the state would be completely delocalized, that is, equally spread over all the atoms. The charges are not equal, in general, and the rms variation measures the localization. Ching and Lin discussed the variation of localization with the position of the state in the valence band, and with the variations in atomic geometry. This analysis was extended by Guttman *et al.* (1980), who summed the fractional charges in the valence band for individual atoms and showed that this leads to net deviations from neutrality for these atoms amounting to about 0.2 of an electronic charge rms. The net charge was strongly correlated with the local deviations of the first and second neighbor distances from their ideal values. The far-infrared absorption by a-Si and a-Ge had already led Klug and Whalley (1976, 1977) to realize that these charges exist, but their conclusion was disputed or not appreciated because it was formulated in terms of atomic displacements from lattice positions. In a later publication (Klug and Whalley, 1982), the integrated IR absorption intensity of a-Si was reinterpreted to yield charges of about 0.23 electrons rms, in remarkable agreement with the theory. A further verification of the predicted charges has been provided by Jeffrey *et al.* (1981), who measured the width of the Si^{29} nuclear magnetic resonance in a-Si. These authors showed that the width could be accounted for nearly quantitatively by a kind of "chemical shift" induced by charges of the calculated magnitude. The fluctuating charge also manifests itself by inducing a fluctuating shift of the Si core states, as has been found very recently by Ley *et al.* (1982). Using high-resolution UV spectroscopy, they have found broadening of the Si 2p levels such as would be produced by static charges of magnitude 0.11 eV rms. The fact that various measures of the charge agree only roughly with each other or with the population analysis is not unexpected, since the latter is not self-consistent, and each method of measurement demands for precise interpretation an analysis of its own, none of which has yet been accomplished.

General discussion of the status of theory is deferred to Section 9.

8. THEORETICAL RESULTS FOR AMORPHOUS HYDROGENATED SILICON

Calculations of the electronic structure of the hydrogenated material are affected by all of the difficulties that attend the theory of pure a-Si, as well as by the serious obstacles posed by our lack of information about the locations of the hydrogen atoms. Nevertheless, a few calculations have been carried out, which will be reviewed below.

Ching *et al.* (1979, 1980) have constructed random-network cluster models in which hydrogen is incorporated in a variety of ways at a single central location and have computed their electronic states by the same method that was used by Ching *et al.* (1977). Mainly on the basis of a comparison of their densities of states with the experimental electron photoemission spectra of von Roedern *et al.* (1977, 1979), they have concluded that mono-, di-, and trihydride units are constituents of a-Si : H, depending on the conditions of preparation. They have also proposed the existence of a "broken bond" model, in which hydrogen atoms are attached to each of a pair of Si atoms that had been first neighbors. The accuracy of the computational scheme seems to have been established by this and earlier applications. However, the conclusions are weakened, in my opinion, by the arbitrariness of the coordinates of the models, which were not always derived either from any assumed but reasonable potential or by minimization of the total electronic energy. It seems also to be generally true that only a single example of each type of system was constructed.

Allan and Joannopoulos (1980, 1981) and Allan *et al.* (1982) have carried out calculations of the densities of states and total energies of various hydride configurations, as well as of other defects in a-Si, using the "Bethe lattice" termination of finite clusters. Interactions with first and second neighbors were included in the empirical Hamiltonian, but these were not otherwise distance dependent, so positional disorder was not taken into account. Naturally, the results then reflect primarily the topology of the models, and their interpretations of the structure seen in the densities of states computed by them, as well as in those computed by Ching *et al.* (1979, 1980), reflect only this aspect.

In the studies of Ching *et al.* (1979, 1980) and of Allan and co-workers, much emphasis is put on the use of the electronic spectrum of a-Si : H to support some structural models and their calculated densities of states in preference to others. This procedure should perhaps be viewed a little skeptically, in the light of the following argument. In a random-network model of a-Si : H with the atoms joined by normal covalent bonds, there are four valence electrons from every Si atom and one from every H atom. Take, say, 100 atoms of which 85 are Si and 15 are H (i.e., a concentration of 15 at. % H, which is in the interesting range). Then of the total of 355 electrons, only 30, or 8.5%, form the Si–H bonds. Therefore, the additional features attributable to the presence of hydrogen can only amount to this fraction of the integrated density of states. Uncertainties in the background level make estimates very difficult, but the spectra published by von Roedern *et al.* (1977, 1979) seem to show very much larger changes. It may be that the soft photoelectrons emitted come mainly from a surface layer that is very rich in hydrogen and that this not only enhances the contribution from the Si–H bonding electrons but also has markedly affected the Si–Si bonds by a change in atomic structure.

Papaconstantopoulos and Economou (1981) have computed the electronic spectrum of a-Si : H using the coherent potential approximation, which required a quasi-crystalline model of the atomic structure.

To complete the personal bias of this section it is necessary to summarize

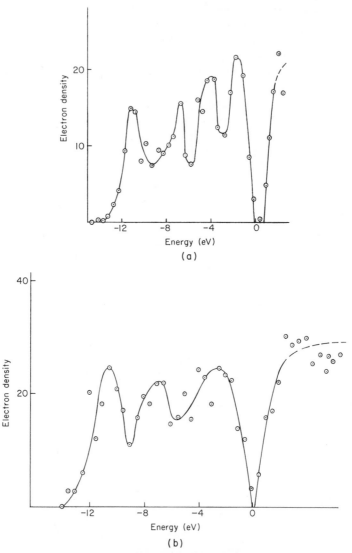

FIG. 5. Local hydrogen densities of states in the valence band, (a) average for three monohydride examples and (b) average for a single dihydride example. (From Guttman and Fong, 1982.)

the results of a recent publication (Guttman and Fong, 1982). In this work little effort has been directed to extracting structural information from the observed spectra. Rather, periodic mono- and dihydride models have been constructed by modification of the models of a-Si referred to earlier. The band structures of these models, which are after all just crystals with large unit cells, have been computed self-consistently using the pseudopotential method in the local-density approximation. The method and the potentials have been tested by others in numerous calculations on a wide range of materials, so that there is little doubt that the accuracy of the results in the present application is limited not by this procedure but by the realism of the models. The effects of both positional and topological disorder are present in the results, but the analysis has been restricted to the former. The main results are the following:

(a)   the total density of states shows, as usual, two broad, unequal maxima in the valence band;

(b)   the density (of states) varies only a few percent from one example to the next;

(c)   the density is much the same whether the repeating unit contains only 30 Si atoms or as many as 54 atoms;

(d)   the gap between occupied and unoccupied states is greater when the mean angular distortion is less, and that in turn is generally reduced by the introduction of hydrogen as monohydride;

(e)   the hydrogen local density of states has four maxima in the valence band for monohydride models but only three for dihydride (see Fig. 5);

(f)   the occupied dihydride states lie on the average about 0.8 eV deeper than the monohydride states; and

(g)   there is the possibility of forming states involving a single H atom and two nearby Si atoms that lie below the bottom of the valence band.

Above all, this work demonstrates that computations of the electronic structure of bulk amorphous hydrogenated silicon are feasible that are comparable in accuracy with those for crystalline semiconductors, and that a lack of structural information is the principal barrier to attaining a deeper understanding of this material.

9. WHAT REMAINS TO BE DONE

While approximate calculations using unrealistic models may still have some utility, it seems clear, in the light of the work described in Section 8, that it is no longer necessary to accept these limitations. There is a need for periodic models with still smaller angular distortions, since those constructed heretofore are not yet fully realistic on the average and the states near the band edges are particularly sensitive to the largest distortions present. Given these improvements, but even without them if necessary, it should be possible to connect the features of the electronic structure of a-Si : H to those of its atomic structure much more fully than at present. This can probably be done more easily using atomic wave functions in the basis, rather than the free electron plane waves used by Guttman and Fong. In this respect, more emphasis should be placed on the topology of the models, since the approximate calculations show its importance.

The fact that a-Si:H is a small-band-gap semiconductor, rather than a metal, in spite of the loss of crystallinity, can be understood in the light of the variety of calculations that have been performed. The ground electronic state of crystalline silicon can be depicted with considerable accuracy as being made up of covalent bonds formed by the superposition of tetrahedrally directed sp^3 hybrid atomic orbitals centered on each pair of Si atoms. The symmetric combination of these orbitals has an electron density maximum midway between the attractive cores and is "bonding"; the antisymmetric combination has zero density at the midpoint and is "antibonding" (i.e., lies much higher in energy). Although these states are broadened into bands in the crystal, no combination of the excited (antibonding) states is possible that lies lower in energy than the uppermost combination of the bonding states, and a gap results. In the amorphous state this picture is still a valid one as long as the first neighbor distances do not deviate too much from that in the crystal, but especially if the interbond *angles* are not so far from the tetrahedral angle that sp^3 hybrids are no longer a good approximation. When the angular distortions become larger, sp^3 hybrids are not good wave functions, the simple bonding–antibonding distinction is blurred, and the gap characterizing a semiconductor decreases and may vanish. This description is supported not only by the tight-binding calculations, in which atomic orbitals form a natural basis, but also by the pseudopotential calculations using a basis of free-electron wave functions.

The concept of fluctuations in atomic charge on Si, which arose naturally in the LCAO calculations as described in Section 7c, is really one that is independent of any model of the structure. The semiquantitative agreement of the theory with three independent experiments should therefore be viewed as an incentive to refine the theory so that the experimental data can be used as probes of the atomic and electronic structure. Since the fluctuation of charge exists only in the presence of disorder, all of its effects are direct measures of the disorder of some kind.

This review has not dealt with a problem of great fundamental interest, namely, the existence of "mobility gaps" and "mobility edges" in disordered materials. It is widely accepted that the one-electron states near the top of the valence band and the bottom of the conduction band in an amorphous semiconductor will show a transition from being localized to being nonlocalized as their energies move away from the center of the energy gap. That is, it is believed that there is a quite sharp change in the character of these states as their energies cross the "mobility edges" and that they contribute to the electrical conductivity by different mechanisms and by very different amounts, depending on whether they lie within the "mobility gap" or outside of it. In principle, this phenomenon can be treated theoretically, provided that the electronic structure can be calculated accurately, and that there exist periodic models of sufficiently different sizes. In the limit of very large numbers of atoms in the repeating unit, the transport properties of the model should approach those of a truly amorphous system, and the periodicity should no longer play a role. Carrying out this ambitious program should stand as a worthy challenge to workers in the field.

REFERENCES

Allan, D. C., and Joannopoulos, J. D. (1980). *Phys. Rev. Lett.* **44**, 43.
Allan, D. C., and Joannopoulos, J. D. (1981). *AIP Conf. Proc.* **73**, 136.
Allan, D. C., Joannopoulos, J. D., and Pollard, W. B. (1982). *Phys. Rev. B* **25**, 1065.
Beeman, D., and Bobbs, B. L. (1975). *Phys. Rev. B* **12**, 1399.
Carlos, W. E., and Taylor, P. C. (1980). *Phys. Rev. Lett.* **45**, 358.
Ching, W. Y., and Lin, C. C. (1978). *Phys. Rev. B* **18**, 2030.
Ching, W. Y., Lin, C. C., and Huber, D. L. (1976). *Phys. Rev. B* **14**, 620.
Ching, W. Y., Lin, C. C., and Guttman, L. (1977). *Phys. Rev. B* **16**, 5488.
Ching, W. Y., Lam, D. J., and Lin, C. C. (1979). *Phys. Rev. Lett.* **42**, 805.
Ching, W. Y., Lam, D. J., and Lin, C. C. (1980). *Phys. Rev. B* **21**, 2378.
Connell, G. A. N., and Temkin, R. J. (1974). *Phys. Rev. B* **9**, 5323.
D'Antonio, P., and Konnert, J. H. (1979). *Phys. Rev. Lett.* **43**, 1161.
Duffy, M. G., Boudreaux, D. S., and Polk, D. E. (1974). *J. Non-Cryst. Solids* **15**, 435.
Etherington, G., Wright, A. C., Wenzel, J. F., Dore, J. D., Clarke, J. H., and Sinclair, R. N. (1982). *J. Non-Cryst. Solids* **48**, 265.
Evans, D. L., Teter, J. P., and Borrelli, N. F. (1974). *AIP Conf. Proc.* **20**, 218.
Guttman, L. (1974). *AIP Conf. Proc.* **20**, 224.
Guttman, L. (1976). *AIP Conf. Proc.* **31**, 268.
Guttman, L. (1981). *Phys. Rev. B* **23**, 1866.
Guttman, L., and Fong, C. Y. (1982). *Phys. Rev. B* **26**, 6756.
Guttman, L., Ching, W. Y., and Rath, J. (1980). *Phys. Rev. Lett.* **44**, 1513.
Henderson, D. (1974). *J. Non-Cryst. Solids* **16**, 317.
Henderson, D., and Herman, F. (1972). *J. Non-Cryst. Solids* **8–10**, 359.
Jeffrey, F. R., Murphy, P. D., and Gerstein, B. C. (1981). *Phys. Rev. B* **23**, 2099.
Joannopoulos, J. D., and Cohen, M. L. (1976). *Solid State Phys.* **31**, 71.
Joannopoulos, J. D., and Yndurain, F. (1974). *Phys. Rev. B* **10**, 5964.
Keating, P. N. (1966a). *Phys. Rev.* **145**, 637.
Keating, P. N. (1966b). *Phys. Rev.* **149**, 674.
Kelly, M. J., and Bullett, D. W. (1976). *J. Non-Cryst. Solids* **21**, 155.
Klug, D. D., and Whalley, E. (1976). *AIP Conf. Proc.* **31**, 229.
Klug, D. D., and Whalley, E. (1977). *Phys. Rev. B* **15**, 209.
Klug, D. D., and Whalley, E. (1982). *Phys. Rev. B* **25**, 5543.
Knights, J. C., and Lujan, R. A. (1979). *Appl. Phys. Lett.* **35**, 244.
Ley, L., Reichardt, J., and Johnson, R. L. (1982). *Phys. Rev. Lett.* **49**, 1664.
Meek, P. E. (1977). *J. Phys. C* **10**, L59.
Moss, S. C., and Graczyk, J. F. (1970). *Proc. Int. Conf. Phys. Semicond., 10th, 1970* p. 658.
Mott, N. F., and Davis, E. A. (1979). "Electronic Processes in Non-Crystalline Materials," 2nd ed., Chapter 7. Oxford Univ. Press (Clarendon), London and New York.

Papaconstantopoulos, D. A., and Economou, E. N. (1981). *Phys. Rev. B* **24,** 7233.

Paul, W. (1980). *Solid State Commun.* **34,** 283.

Polk, D. E. (1971). *J. Non-Cryst. Solids* **5,** 365.

Polk, D. E., and Boudreaux, D. S. (1973). *Phys. Rev. Lett.* **31,** 92.

Postol, T. A., Falco, C. M., Kampwirth, R. T., Schuller, I. K., and Yelon, W. B. (1980). *Phys. Rev. Lett.* **45,** 648.

Rabe, P., Tolkiehn, G., and Werner, A. (1979). *J. Phys. C* **12,** L545.

Reimer, J. A., Vaughan, R. W., and Knights, J. C. (1981). *Solid State Commun.* **37,** 161.

Sadoc, J. F., and Mosseri, R. (1982). *Philos. Mag. [Part] B* **45,** 467.

Spicer, W. E., and Donovan, T. M. (1971). *Phys. Lett. A* **36,** 459.

Steinhardt, P., Alben, R., and Weaire, D. (1974). *J. Non-Cryst. Solids* **15,** 199.

Thorpe, M. F., and Weaire, D. (1974). *Amorphous Liq. Semicond., Proc. Int. Conf., 5th, 1973* p. 917.

von Roedern, B., Ley, L., and Cardona, M. (1977). *Phys. Rev. Lett.* **39,** 1576.

von Roedern, B., Ley, L., Cardona, M., and Smith, F. W. (1979). *Philos. Mag. [Part] B* **40,** 433.

Wagner, C. N. J. (1978). *J. Non-Cryst. Solids* **31,** 1.

Weaire, D., and Wooten, F. (1980). *J. Non-Cryst. Solids* **35–36,** 495.

Weaire, D., Higgins, N., Moore, P., and Marshall, I. (1979). *Philos. Mag. [8]* **40,** 243.

Weinstein, F. C., and Davis, E. A. (1973). *J. Non-Cryst. Solids* **13,** 153.

Zachariasen, W. H. (1932). *J. Am. Chem. Soc.* **54,** 3841.

CHAPTER 12

Experimental Determination of Structure

A. Chenevas-Paule

CENTRE D'ETUDES NUCLÉAIRES DE GRENOBLE
LABORATOIRE D'ELECTRONIQUE ET DE TECHNOLOGIE DE L'INFORMATIQUE
GRENOBLE, FRANCE

I. Introduction

In a disordered system it is impossible to have clear information about the atomic location from the diffraction diagram. In fact, it is necessary to define homogeneity scale for each atom species. In the case of hydrogenated amorphous silicon, it is fundamental to know if hydrogen is distributed homogeneously (alloy model) (Solomon *et al.,* 1978) or clustered in voids (Brodsky, 1980). Hydrogen may also be considered as a probe, if we assume that it decorates defects only, if it allows a study of the a-Si structure. As we shall see later, the problem is complex because very often the growth structure (the so-called columnar structure) is embedded in an intrinsic microstructure. Experiments using nuclear magnetic resonance (NMR) proton have shown that most of the hydrogen is clustered (Reimer *et al.,* 1980). Lamotte *et al.* (1981) using ^{29}Si-NMR has shown directly that hydrogen is not homogeneously distributed. An interesting controversy occurs now about the narrow line of the NMR proton spectra and its very short relaxation time (Carlos and Taylor, 1982; Conradi and Norberg, 1981). But NMR is not capable of giving the homogeneity scale directly. It is vital to determine the homogeneity of the material because the aggregation of one of the species such as hydrogen could be very important on a scale of the order of 10 Å. Beyond this scale one has to consider a multigap system as proposed by Brodsky (1980). The difficulty is that conventional tools for structural investigation (x-rays, electrons) cannot directly see hydrogen.

In this chapter we show that it is possible to prepare homogeneous samples without columnar structure, this kind of structure being due to growth with incomplete coalescence. This homogeneous material produces a typical glasslike fracture and exhibits no resolvable structure. Using high-resolution electron microscopy (HREM), scanning electron microscopy (SEM), and small-angle neutron scattering on these homogeneous samples (a-Si, a-Si: H, a-Si: D), we show that hydrogen seems to be concentrated along linear defects that it decorates. These defects may be disclination cores, a "geometrical locus" of the more disturbed bonds in the material, possibly a linear torsion defect. This kind of conformation favors an electronic structure characteristic of a medium-range order.

1. Structure of Amorphous Silicon — Evolution of Concepts

The word "amorphous" means "structureless," hence to seek a measurement of such a material is a contradiction. The first model for an ideal glass is due to Zachariasen (1932); this model and its famous two-dimensional representation influenced many researchers in this field. It is this model that has lead to the concept of the homogeneous random network widely used for theoretical work on amorphous and glassy semiconductors (Polk and Boudreaux, 1973; Bell and Dean, 1972). From this concept of structure, which takes into account short-range order at the scale of the lattice cells initially considered by Anderson (1958) have evolved models of electronic structure due to Mott, Cohen, and Ovshinsky have appeared (Mott, 1967; Cohen et al., 1969; Davies and Mott, 1970). But homogeneous models fail to describe structural stability as well as physical properties such as conductivity.

Phillips (1979) was the first to take into account the mismatch between the bonding constraints and the number of degrees of freedom in three dimensions and the flexibility required to accommodate the mismatch, and to treat the problem quantitatively. For the a-Si: H case some authors point out that much more hydrogen is present than (Kaplan et al., 1978) is needed to quench the spin resonance signals (Kaplan, 1978). Phillips (1979) proposed a topological model with internal microvoids on the internal surfaces of which dangling bonds reconstruct as they do on crystalline surfaces, giving a global spin density of about 10^{19} cm^{-3}. Hydrogenation occurs by chemisorbtion on the surface of these voids. In fact, many authors have speculated about the existence of voids (D'Antonio and Konnert, 1979) but the shape and the dimensions of these voids were never experimentaly determined.

Knights and Lujan (1979) were the first to observe columnar structures in a-Si : H. Underlying this heterogeneous structure that can also be found in sputtered polycrystalline metals (Thornton, 1974), they assumed that, due to a scale law, structureless materials may be composed of a very thin columnar structure. This structure as proposed by Knights results from growth in which the dependent parameter is the coalescence. This paper (we think) is at the beginning of many other heterogeneous models, such as granular models, two-phase models, and the multiphase model (Paul, 1981).

An increase in the number of phases is perhaps not a good approach in the absence of a realistic Gibbs law for such a system where the number of components is not higher than two. We accept the possibility that some material may be a complex mixture of microcrystals in an amorphous matrix with a columnar structure.

2. NEW THEORETICAL APPROACHES

a. The Rivier-Dufy (1982) Approach

This approach is purely geometrical (or topological) for it consists in demonstrating that odd-numbered rings of bonds (five, seven, . . .), in CRN that are indices of the local curvature, are aligned in such a manner that they surround the core of a disclination (Rivier and Dufy, 1982). Odd-numbered rings can all be stepped through by single lines that avoid any numbered rings and form closed loops or terminate at the surface of the materials. This model takes into account only one kind of disclination.

Disclinations are also defined as rotation defects (Harris, 1977). Rivier (1982) demonstrated that this kind of defect is topologically stable and is suitable case for homotopy. This defect can perhaps also account for some physical properties such as the low-temperature specific heat (tunneling mode). On the other hand Rivier demonstrates that disclinations are the only possible defects in glass. Concerning the homogeneity of the glass it seems to us interesting to cite Rivier (1982) himself: "The glass is treated as a homogeneous continuous substance but unlike a crystal, its homogeneity is not a constructive microscopic symmetry."

b. The Sadoc–Kleman Approach

Starting from curved spaces in order to have the best space filling, this approach consists in searching for the best way to map a regular arrangement in curved spaces into three-dimensional Euclidian space (Kleman and Sadoc, 1979; Sadoc, 1980; Sadoc and Mosseri, 1982). Thus the amorphous disorder appears as a consequence of the mapping of regular structures defined in a three-dimensional curved space. Mapping may induce different kinds of defects; the two-dimensional one works but may be unrealistic with regard to the material connectivity. The disclination created by cutting the structure and adding (or removing) a wedge of material between the two lips of the cut seems to be the best mapping. The final result is similar to those obtained by Rivier concerning the odd-numbered rings of bands and leads to a three-dimensional disclination network.

Recently, Kleman (1983) has proposed for the best final structure two conjugated irregular disclination networks with respectively opposite signs for associated distortions.

If these new descriptions of amorphous material are realistic, because of their high density one must expect single (microscopic) or collective (macroscopic) manifestations of these defects. For a-Si especially, these defects play a very important role in the hydrogenation (effusion–diffusion) mechanisms.

This paper is devoted to showing the existence of these defects by structural studies on posthydrogenated material prepared by sputtering in argon.

II. Electron Microscopy

3. HIGH-RESOLUTION ELECTRON MICROSCOPY (HREM)

High-resolution electron microscopes are now capable of providing unambiguous structural information in crystalline solids at the atomic level (resolution $r \simeq 2$ Å). Thus it is tempting to use this characterization tool to obtain structural information about one-, two-, or three-dimensional defects in a-Si and a-Si : H as was done for crystalline silicon (Bourret, 1972). Nevertheless, it should be noted that the electron scattering coefficient for hydrogen is very low and only important local atomic density fluctuations result in a noticeable contrast. The phase contrast is not well understood, especially since specific defects in amorphous materials are not yet well known (Smith *et al.*, 1980; Stobbs and Smith, 1980). We thus have used this technique essentially to show the existence of some defects. Before describing in detail the experiments, the basic principles of information transfer theory on which HREM is based must be explained (Bourret, 1972).

a. Principles of HREM

The microscope can be considered as an information transfer box (object → image); in which the input signal is altered by distortion and noise. The electron beam extracts information from the electron–matter interaction, but phase and amplitude fluctuations of the incident wave must be expected. Thus one should study the action of the microscope on the electron beam coming out of the object (information transfer) as well as the object action on the incident beam (object transparency).

If we call $\Psi_0(r_0)$ the wave function of electrons when emerging from the object (input signal) and $\Psi_1(r_1)$ the wave function of electrons on the screen (output signal), the microscope provides the transfer from Ψ_0 to Ψ_1. It is necessary to define a transfer function. One can arbitrarily decompose the transfer box as a superposition of three operations:

(1) Fourier transform (FT) giving the spectrum of the input signal $\tilde{\Psi}_0(f)$,

(2) multiplication by a transfer function $T(f)$,

(3) reverse Fourier transform to obtain the output signal

$$\underset{\text{object}}{\Psi_0(r_0)} \xrightarrow{\text{TF}} \underset{\text{diffraction}}{\tilde{\Psi}_0(f) \times T(f)\Psi_1(f)} \xrightarrow{\text{(TF)}^{-1}} \underset{\text{image}}{\Psi_1(r_1).}$$

Therefore if noise is neglected, the input signal can be obtained from $\Psi_1(r_1)$ and $T(f)$.

If $\Psi_0(r_0)$ is the amplitude of the wave coming from the object, $\tilde{\Psi}_0(f)$ is the amplitude of the wave at a point in the focal plane (see Fig. 1). For each frequency of the Fourier transform of $\Psi_0(r_0)$, there is a diffraction in the $\pm\alpha$ direction with $\alpha = \pm\lambda(f)$. Thus the object spectrum may be directly observed in the focal plane of the objective lens. An electronomagnetic objective lens contains an aperture defect. For a tilt with respect to over the axis each ray undergoes a phase displacement of $(2\Pi/\lambda)W(\alpha)$ or $(2\Pi/\lambda)W(f)$ when $|f| = \alpha/\lambda$ (λ is the wavelength of electrons).

Geometrical optics allow us to show that

$$W(f) = \left| \frac{C_s}{4} \lambda^4 f^4 + \frac{\Delta Z}{2} \lambda^2 f^2 \right|, \tag{1}$$

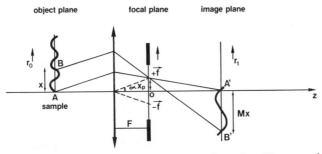

FIG. 1. Schematic diagram for the image-to-object transformation. (Vectors are indicated by overarrows in figures and by bold letters in text.)

where C_s is the spherical aberration coefficient and ΔZ the defocusing value of the focusing defect. We can write the transfer function as follows:

$$T(\mathbf{f}) = \frac{1}{M} \exp\left[\frac{2\Pi i}{\lambda} W(f)\right] B(\mathbf{f}). \tag{2}$$

Where $B(\mathbf{f})$ is the aperture function, the introduction of a diaphragm in the focal plane induces a cutoff for certain spatial frequency ranges. One can also notice that the contrast depends on W and consequently on the defocusing value ΔZ. As will be seen later, ΔZ is a dominant factor in the details that one can observe. Generally speaking, it is always possible to adjust the bandpass to make it as large as possible and choose a good value for ΔZ.

b. Thin Amorphous Films

We will only consider thin amorphous films here because we are just looking at a phase contrast associated with structural defects. For a population of randomly spatially distributed atoms the Fourier transform for the scattering wave is given by

$$F(f) = F_{at}(f) \sum_n \exp(-i r_n \cdot \mathbf{f})$$

for a sample involving only one atom species, which is true for a-Si as well as for a-Si : H,

$$F_H \ll F_{Si} \quad \text{and} \quad n_H < n_{Si},$$

where F_H is the hydrogen electron scattering coefficient and F_{Si} is the silicon electron scattering coefficient.

In the case of a completely random distribution the average value of Σ_n equals 1 and the output frequency of such an object is identical to that of an isolated atom with the intensities being multiplied by the number N of scattering atoms. Such an object confirms the transfer theory very well for it gives high contrast although each atom is a weak object. One can check on thin samples (thickness lower than 100 Å) as was done by Smith et al. (1980) and Bourret (1982) that the microscope acts as a spatial frequency filter the bandpass of which can be varied by varying the defocusing value ΔZ. Thereby a large spatial frequency range can be continuously visualized.

Laser diffraction allows one to determine from the micrographs of the amorphous material the transfer function and the focusing conditions (Bourret, 1982; Smith et al., 1980). It has been shown (Smith et al., 1980; Stobbs and Smith, 1980) that the use of different spatial frequencies does not give more usable structural informations for thicknesses in the 50–100 Å range. But if one wants to work in a differential manner, comparing a-Si and a-Si : H (obtained by posthydrogenation of a-Si), it is necessary to work at a well-defined defocusing value.

Figure 2 gives a good illustration of the resolution of the microscope that we use (JEOL CX 200; $r = 2.2$ Å) by the possibility of measuring directly the interplanar spacing for (111) silicon planes (3.14 Å). It should also be noticed that it is possible to identify in a straightforward manner crystallites as small as 15 Å in diameter in an amorphous silicon matrix.

c. Preparation of Samples and Observation

Samples were made by sputtering in a pure argon atmosphere ($P_{Ar} \simeq 10^{-2}$ Torr) onto fused SiO_2 substrate at 450°C. The substrate (carefully polished) was chosen for its high thermal and chemical stability, to minimize contamination and allow good temperature control at this low pressure (SiO_2 absorbs infrared at $\lambda > 3$ μm), and permit electrical and optical measurements. 100-Å of a-Si was deposited after a long presputtering ($\simeq 1$ hr) to benefit from a gettering effect and to obtain a very pure material, the cathode being made of "FZ" single-crystal silicon.

Some of these films were converted into a-Si : H by plasma posthydrogenation ($P_{H_2} \simeq 10^{-2}$ Torr) at 400°C for 30 min. The a-Si and a-Si : H were treated in pure HF to strip them from their SiO_2 substrates and then placed on special microscope grids prepared to receive very thin films. Figure 3 shows typical micrographs we obtained; Fig. 3a shows a-Si without treatment, Fig. 3b a-Si : H posthydrogenated at 400°C, and Fig. 3c for a-Si : H posthydrogenated at 450°C. These micrographs were obtained under the same defocusing conditions. Micrographs 3b and 3c exhibit a contrast characteristic that is associated with important local density fluctuations of silicon atoms; this kind of contrast has sometimes been interpreted as due to grain boundaries.

To give a realistic interpretation of this contrast we have developed stereoscopy techniques. Figure 4 shows the two micrographs used for this experiment. They have been taken at a lower magnification than that in Fig. 2, tilting the sample by ±10° with respect to the electron beam axis. This stereoscopic analysis shows that the contrast in Fig. 4, (as in micrographs 3b or 3c) is due to cylindrical voids ($\phi \simeq 10$ Å) that are distributed as shown in Fig. 5. Complementary experiments show that these voids have been produced by etching during the strip-off treatment with HF. It should be noted that the diameter of the voids increases with the temperature of the hydrogenation treatment for a given etching time. Pure a-Si samples do not show any contrast, seem to be homogeneously etched, and never show the kind of contrast we see in the hydrogenated samples. We have also prepared pure a-Si film as thin as 50 Å at 450°; they are also homogeneous and exhibit neither a nucleation pattern nor a crystalline phase. From the HREM

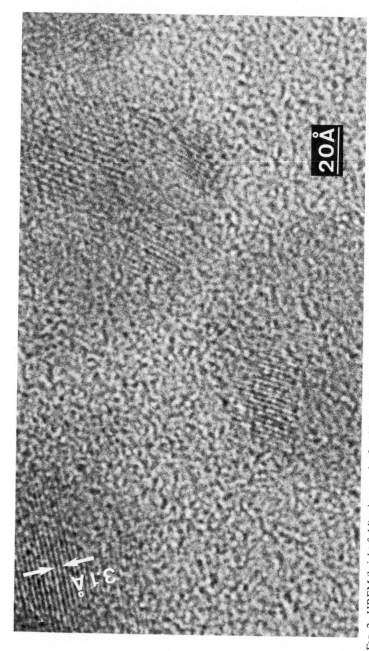

FIG. 2. HREM (bright-field) micrograph of a 100-A a-Si film containing crystalline inclusions having their (111) plane perpendicular to the film plane.

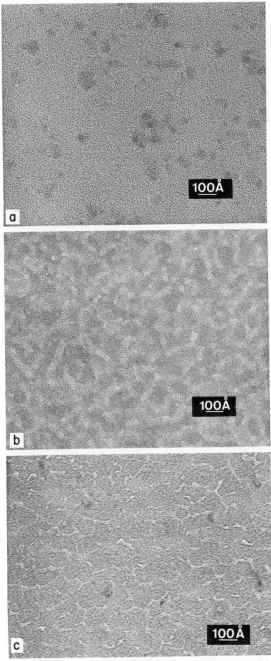

FIG. 3. HREM bright-field micrograph for self-supporting samples: (a) a-Si, (b) a-Si:H posthydrogenated at 400°C, 30 min, (c) a-Si:H posthydrogenated at 450°C 30′.

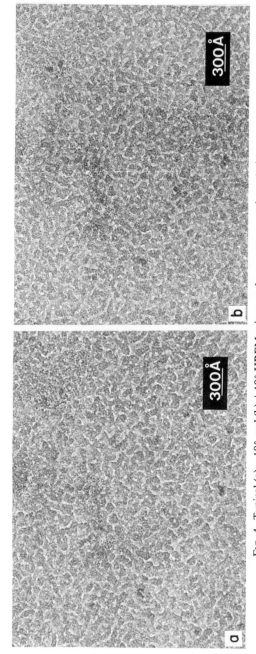

FIG. 4. Typical (a) −10° and (b) +10° HREM micrograph usng stereoscopic analysis.

experimentalist's point of view, they are quite similar to homogeneous thin amorphous carbon films. Consequently, our samples are completely different from those observed by Messier and Ross (1982), which, because of growth and incomplete coalescence, exhibit a super network of voids connected to a so-called columnar structure. In addition micrographs taken after a-Si thin films (50–100 Å thick) on these special HREM grids were submitted to the plasma hydrogenation treatment do not exhibit any density fluctuation or any modification compared to the initial nonhydrogenated state.

To interpret these experimental facts we propose the following processes for hydrogenation:

1. During the plasma hydrogenation treatment, hydrogen diffuses into the material "via" linear defects that it decorates. Such defects were proposed by many authors to be disclinations (Rivier, 1979; Sadoc and Mosseri, 1982) in an attempt to explain some universal properties of noncrystalline materials (low temperature, specific heat, plasticity, etc.).

2. This inhomogeneous distribution of hydrogen in the material induces large local fluctuations of the chemical potential, which are revealed by pure HF acting as a specific etchant of hydrogen-rich regions. These etch pits do not form in pure a-Si because chemical potential fluctuations associated with these linear defects are too weak when undecorated.

It is important to notice that this microcorrosion phenomenon is not specific to amorphous materials; the HREM experimentalists always observed pinholes in crystalline thin films along decorated dislocations (A. Bourret, private communication).

Figure 5 deserves another comment because this figure indicates that a certain axial anisotropy exists in the distribution of the filaments. This is not

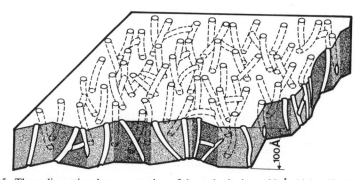

FIG. 5. Three-dimensional representation of the etch pits in a 100-Å-thick a-Si:H sample.

true, and it is probably due to the fact that only those defects that emerge at the surface may be etched. Hence it was also possible to explain the relatively low density of defects found by this method compared to the theoretically predicated density of disclination in amorphous solids.

In conclusion, HREM is not yet able to give straightforward evidence on the structural defect in amorphous materials (Bourret, 1982; Smith *et al.,* 1980). It is probably necessary to work on thinner films ($\simeq 30$ Å). But it seems that by combining HREM, decoration, and corrosion (etch pits) one should be able to obtain valuable information about defects in a-Si.

4. SCANNING ELECTRON MICROSCOPY (SEM)

This study is part of a preliminary and elementary characterization of samples used for neutron scattering experiments. The samples were prepared by RF sputtering onto fused SiO_2 substrates (0.5 mm in thickness). The temperature of the substrate heater was 400°C. The sputtering was argon or argon plus hydrogen, or deuterium ($P_{total} \simeq 10^{-2}$ Torr). The thickness of these samples was between 10 and 30 μm. We have compared this kind of sample with those sputtered onto crystalline silicon at room temperature. Figures 6a and 6b show the scanning electron micrographs of fractured surfaces of thick films parallel to the growth direction for room and high-temperature material. Similar results were obtained by Knights and Lujan (1979) using the glow-discharge technique. On high-temperature samples (Fig. 6b) there is no resolvable structure even at 80,000 magnification (using a JEOL JSM 85 SEM), the limit of resolution being 60 Å. It should be noted also that the fractured surface of Fig. 6b is very smooth and characteristically glasslike (conchoidal fracture). Inspired by HREM experiments we have studied chemical etching of these fractures in order to reveal some possible unresolved structure, a columnar one, as presumed by Knights and Lujan (1979) in an apparently homogeneous material. We used 20% KOH solution in water at 70°C because this reagent is known to be highly anisotropic. A similar technique was used by Messier *et al.* (1980) and Ross and Messier (1981) for revealing columnar structure but they used a HNO_3–HF reagent. Here we shall just look at a high-temperature hydrogenated material typically illustrated by Fig. 6b. Samples were stripped from their substrates by pure HF treatment, fractured, and etched. Figures 7a, 7b, 7c, and 7d illustrate typical etch patterns obtained on this kind of material. In Fig. 7a we can see etched and unetched faces: the unetched face is obtained by fracture; after etching, the etched face shows stratification patterns due perhaps to small fluctuations of some preparation parameters during the growth, which lasts several hours. In Fig. 7b an etched film on its substrate is shown. Etch temperature was 80°C. The etch pits have a hemispherical shape. Figures 7c and 7d were obtained after a 55°C etching.

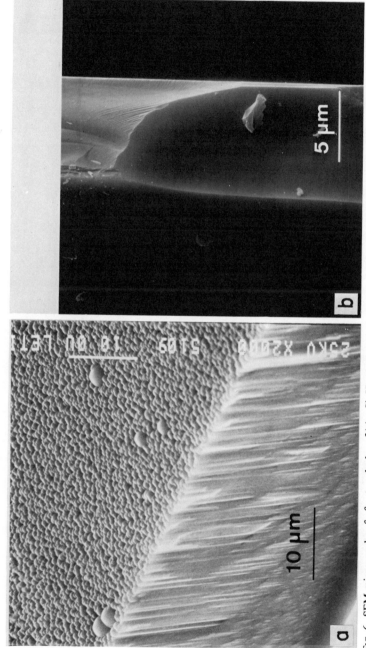

FIG. 6. SEM micrographs of a fractured edge of (a) a-Si:H sample sputtered onto a low-temperature substrate of SiO_2, (b) a-Si:H sputtered onto a high-temperature (400 °C) SiO_2 substrate.

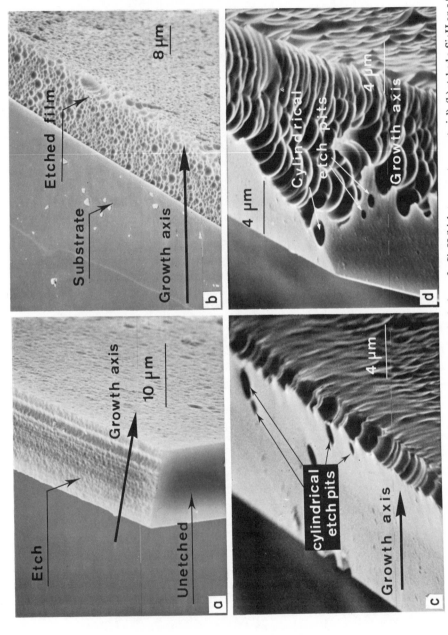

FIG. 7. SEM micrographs of a fractured edge of (a) etched and unetched edges of a-Si : H (high-temperature material), (b) etched a-Si : H on a SiO$_2$ substrate, (c), (d) typical etch pits obtained with 60°C KOH reagent.

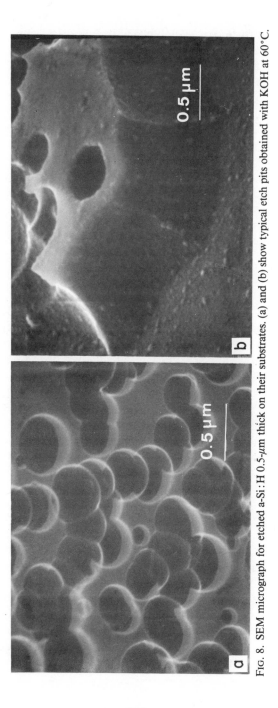

FIG. 8. SEM micrograph for etched a-Si:H 0.5-μm thick on their substrates. (a) and (b) show typical etch pits obtained with KOH at 60°C.

261

Near this temperature typical etch pits with cylindrical symmetry (perpendicular or parallel to the growth direction) were obtained. Under the same conditions, a-Si without hydrogen is etched uniformly. Similar patterns are also obtained on thinner samples (≈ 0.5 μm thick) by the same chemical treatment, as shown in Figs. 8a and 8b, their axis being perpendicular to the film. The symmetry of these etch pits is remarkable, as is also the fact that the columnar structure was never revealed in the high-temperature material. So we think we have demonstrated that it is possible to have films without incomplete coalescence figures (columnar structure), and therefore homogeneous growth without nucleation. Such samples are suitable for the microscopic study of the hydrogen distribution and atomic structure of a-Si:H. Our conclusions for this section are the same as for the HREM section regarding the homogeneity of the material. For a-Si:H growth it seems that two extreme modes exist:

Filamentary growth (Tatsumi *et al.,* 1979)
Homogeneous growth without nucleation

d. Remarks About Etching Patterns

We have already used several times the word "disclination." This word corresponds to a linear torsion defect present in condensed matter such as liquid crystals and cylindrical crystals, already described in a general manner by Harris (1977). The theory of etch pit formation in crystals (Heimann, 1982) is very elaborate, so we could not think of approaching our problem from this angle. Nevertheless, some analogy exists between dislocations and disclinations: they are linear defects inducing elastic stresses and having core energies. The scale of defects revealed in this chapter is very different from those obtained in the HREM chapter. This is not surprising in fact because the reagents are not the same so their selectivities differ. In addition it is necessary to take into account the linear extension of the defects as well as their emergence at the surface. There are also at least two kinds of disclinations obtained from the mapping of a positively or negatively curved space (Kleman, 1983): straight segments and rings that do not emerge.

Heimann (1982) relates that in 1820 mineralogists used chemical attack in order to identify crystallographic symmetries of natural minerals in the pre-x-rays days. In 1949 the relationship between etch pits and dislocations was established irrefutably. We hope that the relation between the etch pits we observed and disclinations will be discovered more quickly. It must also be noted that these experiments are very simple and should have a new resurgence due to the availability of high-resolution SEM with field emission filaments. But finding the adequate specific reagent lies between art and science.

III. Small-Angle Neutron Scattering

5. GENERALITIES

Let us first recall some characteristics of neutron scattering to understand the usefulness of small-angle neutron scattering (SANS) to the study of the system (H, D, Si).

Neutrons allow the study of physical chemistry phenomena on a time scale in the range $10^{-8} - 10^{-14}$ sec and a spatial scale in the range $0.1 - 1000$ Å. This is due to the fact that these non-"relativistic" particles for energies $E = \frac{1}{2}mV^2$ of a few kT are associated with wave lengths $\lambda = h/mV$ of a few angstroms. For similar wavelengths electrons or photons have energies respectively 10^4 or 10^5 times higher. The neutron – matter interaction occurs essentially through the atomic core, and it should be noted that neutrons have a $\frac{1}{2}$ spin. Atomic cores are characterized with respect to this interaction by a scattering length (b) and the fact that they can scatter neutrons in either a coherent or incoherent manner, depending on the nature of the atoms. In our case only Si, D, and H are studied and Table I gives their characteristics.

In this part we shall not analyze the incoherent part, which is isotropic and consequently gives a weak contribution to the scattered intensity at small angles. The analysis of this incoherent part taken over 4Π steradians might give information on the dynamics of scattering centers and lead to a genuine vibrational spectroscopy. For a careful interpretation of SANS experiments it is necessary to subtract this isotropic part from the total intensity.

6. SANS EXPERIMENTS

Except for the energy and wavelength of neutrons, SANS experiments are quite similar to other low-angle scattering experiments (e.g., x-rays and electrons). For our system the main difference between different techniques is the very large σ_{coh} for hydrogen with respect to neutrons (Kostorz, 1979; Roth *et al.*, 1974; Bellissent, 1981; Bellissent *et al.*, 1982). Small-angle scattering is induced by every kind of inhomogeneity or imperfection

TABLE I

	$\langle b \rangle \times 10^{-15}$ m	$\sigma_{coh} \times 10^{-24}$ cm^2	$\sigma_{inc} \times 10^{-24}$ cm^2
H	-3.74	1.8	81.67
D	6.67	5.6	2.
Si	4.14	2.15	0.1
	$\sigma_{coh} = 4\Pi b^2$		

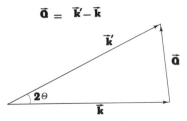

$$\bar{\mathbf{q}} = \bar{\mathbf{k}}' - \bar{\mathbf{k}}$$

FIG. 9. Scattering diagram.

enclosed in the sample, the dimensions of which are in the 10–1000 Å range. The measurement of the scattering intensity at small angles as a function of the angle θ of incidence yields information about the shape and size of these inhomogeneities and the spatial distribution of hydrogen. In fact, the small-angle scattering phenomenon is not exactly defined by an angle but rather by the scattering vector $|\mathbf{Q}| = (4\Pi/\lambda) \sin \theta$.

For elastic scattering $\mathbf{Q}$ is defined as the difference between the incident wave vector $\mathbf{k}$ and the scattering wave vector $\mathbf{k}'$ (Fig. 9). It should be noted

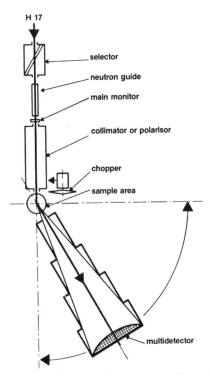

FIG. 10. Experimental diffractor D17 SANS setup. (From Maier, 1981).

TABLE II

INSTRUMENT DETAILS

Beam tube	H17 cold neutron guide
Monochromation	Velocity selectors
Detector	Planar square multidetector with a linear 0.5-cm resolution in both dimensions
Sample detector distance	0.8 m, 1.4 m, or 2.8 m
Detector rotation around the sample (in the horizontal plane)	0–90°
Wavelength range	8–20 Å
Wavelength spread (FWHH) $\Delta\lambda/\lambda$	10 or 5%
Minimum beam-stop size in front of the detector	$4 \times 6 \text{ cm}^2$
q range	5.10^{-3}–1 Å^{-1}
Beam flux at the sample	10^6 n^{-1} sec^{-1} cm^{-2} at 12 Å (see Fig. 10)
Typical sample size (i.e., sample cross-section area perpendicular to the incident beam)	$1 \times 1 \text{ cm}^2$
Minimum background	3–5 n sec^{-1} for the whole multidetector

that the properties of Fourier transforms, measurements at low $\mathbf{Q}$ do not allow resolution of structural details over distances smaller than $d_{min} \simeq \Pi/|\mathbf{Q}|_{max}$, $\mathbf{Q}_{max}$ being the maximum value accessible in the experiment (Kostorz, 1979). In our experiments we have used the D_{17} diffractometer from ILL Grenoble (Fig. 10); the characteristics of this spectrometer are summarized in Table II (Maier, 1981).

D_{17} (Maier, 1981) is a two-axis spectrometer with a multidetector for experiments in which the scattering vector $|\mathbf{Q}|$ ranges from 8.10^{-3} to 0.3 Å^{-1} requiring high resolution in $\mathbf{Q}$ space. It is thus designed for studying high-order diffraction peaks of large periodic structures (low resolution in real space). It can be operated as a classical small-angle scattering spectrometer or as a single-crystal diffractometer.

D_{17} is equipped with a LETI BF$_3$-multidetector, 64×64 cm^2 in area with 16,000 counting cells each 5×5 mm^2, in a planar square matrix arrangement. This detector can be positioned 0.8, 1.4, or 2.8 m from the sample and can be rotated around the sample 0 to 90° with respect to the incident beam direction. The incident beam as well as the scattered beam are in vacuum.

The apparatus is installed at the exit of the cold neutron guide H17. The range of available neutron wavelengths is restricted to 8–10 Å. The beam monochromation is achieved by velocity selectors giving a triangular wavelength distribution with a full width a half maximum (FWHH) $\Delta\lambda/\lambda$ of 5–10%.

For classical small-angle scattering the q resolution, Δq, of the apparatus is, to a first approximation, simply proportional to $|\mathbf{Q}|\,\Delta\lambda/\lambda$ at high scattering angles. At low angles, $\mathbf{Q}$ depends on the collimation of the incident beam. A lowest reasonable limit is 2×10^{-3} Å^{-1}. The divergence of the incident beam can be varied between 2.10^{-2} rad and 5.10^{-3} rad.

Roughly speaking, the scattering intensity may be written as follows:

$$I = I_{inc} + I_{coh}. \tag{3}$$

I_{inc} the incoherent scattered intensity is isotropic although this part is very large due to the very high σ_{inc} of hydrogen; only a small portion of this kind of neutron is taken into account in the small-angle detection area. In addition complementary experiments allow subtracting this scattered part, as well as performing useful normalizations (we will not discuss these techniques here).

If the scattering is due to the presence of low-density clusters in a higher-density matrix, $I_{coh}(Q)$ may be written (Cohen and Ovshinsky, 1969) for $|\mathbf{Q}| \rightarrow 0$ (Bellissent et al., 1982).

$$I_{coh}(Q = 0) \propto \rho_c V_c^2(\xi_c - \xi_m)^2, \tag{4}$$

where ρ_c is the cluster density and V_c the cluster volume fraction.

$$\xi = \sum_c x_{i-}^c b_i^c; \qquad \xi m = \sum_j x_j^m b_j^m.$$

x_i, x_j represent the concentrations, b_i, b_j being the scattering lengths for the different atomic species that constitute the two "phases." One can see that Eq. (4) is too aggregate to give a hint of the cluster composition. The isotopic substitution for one component (H $\rightarrow$ D in this case) allows separating the contributions of x_i^c and x_j^m. Also when the material becomes homogeneous, the scattered contribution I_{coh} (Eq. 4) goes to zero.

Among many models for the determination of the shape of inhomogeneities in a sample from $I(Q)$, the Guinier law gives a good preliminary approach. This law may be written as follows:

$$I(Q) = I(Q = 0) \exp(-\tfrac{1}{3}Q^2 R_g^2), \tag{5}$$

where R_g is the gyration radius, a general concept applicable to all particles and defined by (Kostorz, 1979):

$$R_g^2 = \frac{1}{V_p} \int_{V_p} r_g^2 q(r_g)\, dr_g, \tag{6}$$

where V_p is the volume of the particle, $q(r_g)$ is the geometrical cross section of the particle along a plane normal to a direction **g**, and r_g a distance from the origin inside the particle **g** is a direction in the plane perpendicular to the incident wave vector **k**. The Guinier approximation is valid only for dilute uncorrelated particles.

7. EXPERIMENTAL RESULTS

Scattering experiments (Bellissent *et al.*, 1982) were conducted on four samples (a-Si, a-Si:H a-Si:D.₅H.₅ and a-Si:D); the H or D content is around 15 at. %. The samples were the same as those described in Section 4 and were prepared at a relatively high temperature, $T_s \simeq 400°C$, onto fused silica. The thickness of the amorphous material was around 30 μm and four stacked samples were used to obtain an adequate signal-to-noise ratio. By changing the sample–detector distance it was possible to attain momentum transfer values between 5.10^{-3} and $2.8 \ 10^{-1} \ \text{Å}^{-1}$.

Figure 11 shows the scattered intensities obtained versus Q directly on a

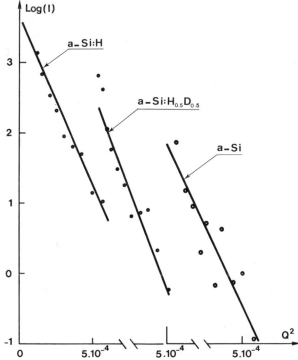

FIG. 11. Guinier plots of the scattered intensity for a-Si, a-Si:H, a-Si:D.₅ H.₅. (From Bellissent *et al.*, 1982.)

Guinier plot. The scattered intensity for the deuterated sample is not shown because it is too low, which is in good agreement with relations (4) and (5) and the values of Table I. Experimental results summarized in Fig. 11 call for the following comments:

1. All samples exhibit inhomogeneities that can be characterized by a 120-Å radius of gyration deduced from the Guinier plot. This implies that inhomogeneities are already present in the sample without hydrogen.

2. The hydrogen or deuterium content affects only the value of $I(Q)$ and not the radius of gyration. To a first approximation, $I(Q = 0)$ obeys relation (4); that is, fluctuations of the intensity are purely due to isotropic substitution, as shown by the $I(Q)$ for a-Si:$H_{.5}D_{.5}$.

From the above two comments, it follows that hydrogen obviously decorates preexisting defects in a-Si. In addition, one can see in Fig. 12, which is the experimental plot of $I(Q)$ for a-Si and a-Si:H, a low intensity ring near $Q = 0.22$ Å^{-1}. This ring may be related to a correlation length (L_c) of about 30 Å (L_c is approximately the distance between heterogeneities).

From SANS experiments we have obtained two characteristic lengths $R_g \simeq 120$ Å and $L_c \simeq 30$ Å. Perhaps it is too early to give an *exact* description for these objects (shape and spatial distribution) in a-Si and a-Si:H. Nevertheless the more realistic if not the best picture that we may have about the distribution of hydrogen from these characteristic lengths is the following: (1) H is localized along filaments whose length is greater than 120 Å; and (2) The average distance between these filaments is 30 Å.

In Fig. 12 we can see also that for large $Q(Q > 3 \ 10^{-2}$ Å$^{-1})$, $I(Q)$ increases with $H(\Delta I(Q) \simeq 5\%)$; we have calculated that this increase is due to the incoherent scattering part of hydrogen. To a first approximation the scattered coherent part may be neglected; thus the fraction of hydrogen that is homogeneously distributed is very small.

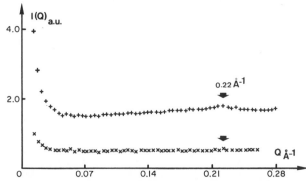

FIG. 12. Scattered intensity for a-Si:H(+) and a-Si(×) versus |Q|. (From Bellissent *et al.*, 1982.)

As we said previously the Guinier law is correct only for particles in low concentration in a matrix. Our case ($L_c \simeq 30$ Å) is not very suitable for this law, so we have to use other more realistic models for a finer analysis.

IV. Conclusion

In conclusion we have shown that it is possible to prepare a-Si:H where structural defects due to incomplete coalescence (columnar structure) are absent. In this kind of homogeneous amorphous material hydrogen clusters along a filamentary network, the distance between these linear defects being about 30 Å. These filamentary lines are topologically consistent with defects theoretically foreseen by Kleman, Sadoc, and Rivier (Rivier, 1979, 1982; Rivier and Dufy, 1982; Sadoc, 1980; Sadoc and Mosseri, 1982; Klein and Sadoc, 1979; Kleman, 1983). These observations are also consistent with the new model given by Kleman (1983). From a topological point of view it should be noticed that this structure allows an optimum connectivity for the material as well as for the defects (linear voids); this condition was previously anticipated by Phillips (1979). The consequences of such a structure may be summarized as follows:

1. Hydrogen decorates preexisting linear defects.

2. The electronic structure of a s-Si:H may be better described by a Brodsky-like quantum well system containing pure but noncrystalline silicon regions without hydrogen rather than by an alloy or a two-phase system. The Brodsky description may be applied to the pure regions (Brodsky and Di Vincenzo, 1982).

3. Linear defects (disclinations), hydrogenated or not, are the source of elastic and electrostatic microscopic band-gap fluctuations.

4. The interaction between impurities, doping atoms, point defects in general, and linear defects (sources of elastic perturbations) must be considered. They are probably the origin of the doping inefficiency.

5. Two kind of electronic defects may exist, one related to disclinations and responsible for the tail states, the other consisting of point defects located in the quasi-pure silicon region and having the same character as deep centers in crystalline semiconductors.

Nevertheless, a lot of work has yet to be done to validate this model. Concerning neutrons, lower $|Q|$ experiments are needed to determine more precisely the chemical composition around the linear defects. This may be done by using careful measurements near $|Q| = 0$ on a-Si, a-Si:H, a-Si:H,D samples including for a-Si:H,D the zero mixture. The small scattering ring we have observed at a 0.22 Å^{-1} transfer momentum should be analyzed with a better signal-to-noise ratio. In addition, other representations than

Guinier's remain to be found. Work needs to be done on the different models that we can deduce from HREM or SEM, etch pits observations. Finding the better representation also requires more precise information on the state of hydrogen in the material (molecular hydrogen, kind of bond, etc.), a microscopic model for hydrogenation of different kinds of disclinations, and an estimate of the ratio of ring disclination to linear disclination.

We also need large $|Q|$ neutron diffraction experiments for short-range order determinations. The idea is therefore to obtain the partial structure factors and in particular the $Si-H$, $Si-Si$, and $H-H$ partials by the isotopic substitution technique comprising the preparation of the zero mixture (H/D) for which the coherent scattering by H and D component is zero. Molecular hydrogen could also be detected during this experiment. Such an attempt has already been made by Leadbetter *et al.* (1980) but failed owing to lack of deuterated samples and problems with data analysis. EXAFS, which does not see hydrogen, may be an excellent complementary means to investigate the influence of hydrogen on short-range order. It should be noted that neutrons as well as NMR and x-rays require large amounts of material, in contrast with the very thin films used in HREM (50–100 Å). If complete coalescence phenomena are avoided, we think that it is possible to compare thick and thin material just as can be done with crystals. The only way to visualize disclinations remains electron microscopy correlation with chemical etching. Much work remains to be done in this area.

For identification of disclination we must know the sign of the strain and a way to determine it. Finally we have shown that hydrogen is located in the linear defects, so we have now to understand how these defects are microscopically hydrogenated and what the consequences are of hydrogenation for the microscopic strains around the core of these defects. Theoretical and experimental work on this subject now brought together must be reconciled.

ACKNOWLEDGMENTS

This work was supported in part by the French Agency for the Mastery of Energy (AFME). The author gratefully acknowledges A. Bourret and R. Bellissent for many very fruitful discussions on electron microscopy and neutron scattering, respectively. Samples were carefully grown by R. Cuchet.

REFERENCES

Anderson, P. W. (1958). *Phys. Rev.* **5,** 1492.
Bell, R. J., and Dean, P. (1972). *Philos Mag.* [8] **25,** 1381.
Bellissent, R. (1981). Thesis, Université Pierre et Marie Curie, Paris.
Bellissent, R, Chenevas-Paule, A., and Roth, M. (1983). *Physica* **117–118B,** 941.
Bourret, A. (1972). "La microscopie électronique à très haute résolution". Ecole d'Été de Villard de Lans.

Bourret, A. (1982). HREM School, Arizona State University, Tempe.

Brodsky, M. H. (1980). *Solid Stte Commun.* **36,** 55.

Brodsky, M. H., and Di Vincenzo, D. P. (1983). *Physics* **117–118B,** 971.

Carlos, W. E., and Taylor, P. C. (1982). *Phys. Rev. B* **26,** 3605.

Cohen, M. H., Fritzsche, H., and Ovshinsky, S. R. (1969). *Phys. Rev. Lett.* **22,** 1065.

Conradi, M., and Norberg, R. (1981). *Phys. Rev. B* **25,** 2285.

D'Antonio, P., and Konnert, J. H. (1979). *Phys. Rev. Lett.,* **43,** 1161.

Davies, E. A., and Mott, N. F. (1970). *Philos. Mag.* [8] **22,** 903.

Harris, W. F. (1977). *Sci. Am.* **237,** 130.

Heimann, R. B. (1982). *Cryst: Growth, Prop., Appl.* **8,** 173.

Kaplan, D., Sol, N., and Velasco, G. (1978). *Appl. Phys. Lett.* **33,** 440.

Kleman, M. (1983). *J. Phys. Lett.* **44,** L295.

Kleman, M., and Sadoc, J. F. (1979). *J. Phys. Lett.* **40,** L569.

Knights, J. C., and Lujan, R. A. (1979). *Appl. Phys. Lett.* **35,** 244.

Kostorz, G., ed. (1979). "Treatise on Materials Science and Technology," Vol. 15, p. 227. Academic Press, New York.

Lamotte, B., Rousseau, A., and Chenevas-Paule, A. (1981). *J. Physique Suppl.* **42,** 839.

Leadbetter, A. J., Rashid, A. A. M., Colenutt, N., Wright, A. F., and Knights, J. C. (1980). *Solid State Commun.* **33,** 973.

Maier, B. (1981). "Neutron Beam Facilities Available for Users." Grenoble Cedex, France.

Messier, R., and Ross, R. C. (1982). *J. Appl. Phys.* **53,** 6220.

Messier, R., Krishnaswamy, S. V., Gilbert, L. R., and Swab, J. (1980). *J. Appl. Phys.* **51,** 1611.

Mott, N. F. (1967). *Adv. Phys.* **16,** 49.

Paul, W. (1981). *Amorphous Liq. Semicond., Proc. Int. Conf., 9th, 1981* p. 1165.

Phillips, J. C. (1979). *Phys. Rev. Lett.* **42,** 1151.

Polk, D. E., and Boudreaux, D. S. (1973). *Phys. Rev. Lett.* **31,** 921.

Reimer, J. A., Vaughan, R. W., and Knights, J. C. (1980). *Phys. Rev. Lett.* **44,** 183.

Rivier, N. (1979). *Philos. Mag.* [8] **40,** 859.

Rivier, N. (1982). *Philos. Mag.* [8] **45,** 1081.

Rivier, N., and Dufy, D. M. (1982). *J. Physique* **43,** 293.

Ross, R. C., and Messier, R. (1981). *J. Appl. Phys.* **52,** 5329.

Roth, M., Cotton, J. P., and Ober, R. (1974). "Introduction à la spectrométrie neutronique. Troisième partie: Diffusion aux petits angles." Laboratoire Léon Brillouin, Gif/Yvette, France.

Sadoc, J. F. (1980). *J. Non-Cryst. Solids,* **44,** 1.

Sadoc, J. F., and Mosseri, R. (1982). *Philos. Mag.* [8] **45,** 467.

Smith, D. J., Saxton, W. O., Cleaver, J. R. A., and Catto, C. J. D. (1980). *J. Microsc. (Oxford)* **119,** 19.

Solomon, I., Perrin, J., and Bourdon, B. (1978). *Proc. Int. Conf. Phys. Semicond., 14th, 1978* p. 689.

Stobbs, W. M., and Smith, D. J. (1980). *J. Microsc. (Oxford)* **119,** 29.

Tatsumi, Y., Hirata, M., and Shigi, M. (1979). *Jpn. J. Appl. Phys.* **12,** 2199.

Thorton, J. A. (1974). *J. Vac. Sci. Technol.* **11,** 666.

Zachariasen, W. H. (1932). *J. Am. Chem. Soc.* **54,** 3841.

CHAPTER 13

Pressure Effects on the Local Atomic Structure

S. Minomura

INSTITUTE FOR SOLID STATE PHYSICS
UNIVERSITY OF TOKYO-ROPPONGI
TOKYO, JAPAN

I. Introduction

In recent years there has been a rapidly growing interest in various photovoltaic applications of hydrogenated amorphous silicon (a-Si:H). The electronic properties of plasma-deposited a-Si:H films are mostly determined by the gap states, which are very sensitive to the atomic structure and defects. The information about the local order of a-Si:H films has been derived from a variety of measurements: vibrational spectroscopy, x-ray (also electron or neutron) diffraction, EXAFS, photoemission spectroscopy, ESR, NMR, and so on. The infrared spectra are identified as four types of Si–H bonding conformations: SiH, SiH_2, SiH_3, and $(SiH_2)_n$-complex (Brodsky *et al.,* 1977; Lucovsky *et al.,* 1979). The Si–H bonding is formed predominantly in the films deposited from the glow-discharge plasma of silane at high substrate temperatures of about 250°C. Weaire and associates (1979) have suggested that the H atoms are grouped in clusters rather than dispersed throughout the structure. On the other hand, in the reactively sputtered films the H atoms are incorporated as a mixture of Si–H and Si–H_2 bonding. Moreover, the sputtered films contain Ar atoms (Tanaka *et al.,* 1980). The alloying effect of H atoms gives rise to a reduction of the large number of gap states and an increase of the optical energy gap.

Plasma-deposited a-Si:H films demonstrate the systematic changes in local order and gap states with H concentration, C_H, which depend critically

273

on the preparation method and preparation condition. The glow-discharge plasma of silane has been studied for understanding the nucleation and growth kinetics of a-Si:H films (Matsuda and Tanaka, 1982).† The plasma contains about 10^{10} electrons cm^{-3} with energies of $1-10$ eV. The emission spectra are identified in terms of primary ions produced by electron impact and secondary ions produced by ion–molecule reactions. The growth rate of a-Si:H films, however, depends on the neutral radicals rather than ions. The nucleation and growth are strongly influenced by electron and ion bombardment.

High-pressure research plays an important role for understanding the structural and electronic properties of condensed matter. One can use high pressure as a macroscopic parameter to change the volume and electronic states in a controlled way. As a consequence, the application of pressure can provide a new ground state with different electronic properties. The pressure-induced effects reflect the bonding nature and the dimensionality of the network.

It is the purpose of this review to discuss the experimental results on the pressure-induced effects of local order of a-Si:H films and the pressure-induced phase transitions from semiconductor to metal. The plan of discussion is as follows: The first section describes high-pressure experimental techniques. The second section deals with the local order derived from the disorder-induced Raman spectra. The third section discusses the local order in terms of x-ray diffraction intensity profiles. The fourth section describes pressure-induced phase transitions and the associated changes in optical energy gap, electrical resistivity, and atomic structure.

II. Experimental Techniques

Pressure-induced effects and phase transitions in plasma-deposited a-Si:H films have been studied by measurements of Raman scattering, x-ray diffraction, fundamental optical absorption, and electrical resistivity, using opposed anvil pressure cells. The diamond-anvil pressure cell is becoming the most powerful device for high-pressure spectroscopy and x-ray diffractometry. A cross-sectional drawing of the diamond-anvil cell driven by a piston and screw device (Takemura *et al.*, 1979) is shown in Fig. 1. The diamond anvils are aligned by adjusting the tilting and translating screws from outside under a microscope. This device is capable of generating pressures up to 500 kbar with a metal gasket and a pressure medium. The pressure is determined from the pressure dependence of ruby fluorescence lines (Piermarini *et al.*, 1975).

---

† See Uchida, Chapter 3, this volume.

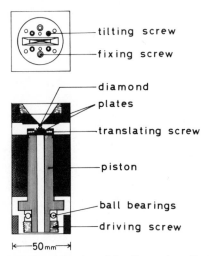

tilting screw

fixing screw

diamond

plates

translating screw

piston

ball bearings

driving screw

|←—50 mm—→|

FIG. 1. Cross-sectional drawing of the diamond-anvil pressure cell.

A schematic diagram for measuring Raman scattering in the diamond-anvil cell (Ishidate *et al.*, 1982) is shown in Fig. 2. The laser beam from an Ar or Kr ion laser is focused to a fine spot on the sample inside the cell by the exciting system. The backward-scattered light is focused onto the slit of a double monochrometer by the collecting system. The Raman signal is detected by photon counting and stored in a multichannel analyzer.

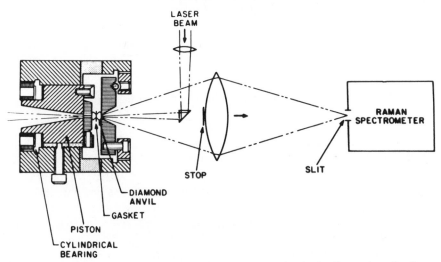

FIG. 2. Schematic diagram for measuring Raman scattering in the diamond-anvil cell.

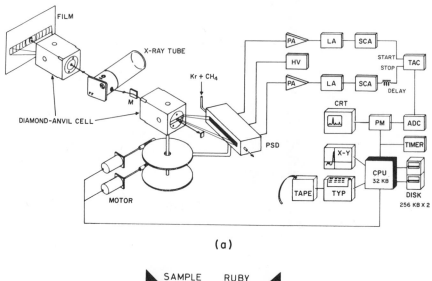

(a)

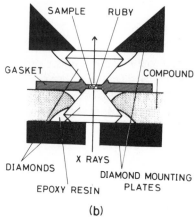

(b)

FIG. 3. Schematic diagram for measuring x-ray diffraction in the diamond-anvil cell: (a) general arrangement, (b) sample assembly.

A schematic diagram for measuring x-ray diffraction intensity in the diamond-anvil cell (Fujii *et al.*, 1980) is shown in Fig. 3. The incident x-ray beam from a molybdenum target is monochromatized with pyrolytic graphite and adjusted to a fine spot on the sample by rotating a doubly eccentric collimator. The diffraction intensities are detected by a position-sensitive proportional counter (PSPC). The PSPC can cover a scattering-angle range of 22.6° with an angular resolution of 0.038° in a single measurement. This device is capable of providing rapid data collection with a high signal-to-noise ratio.

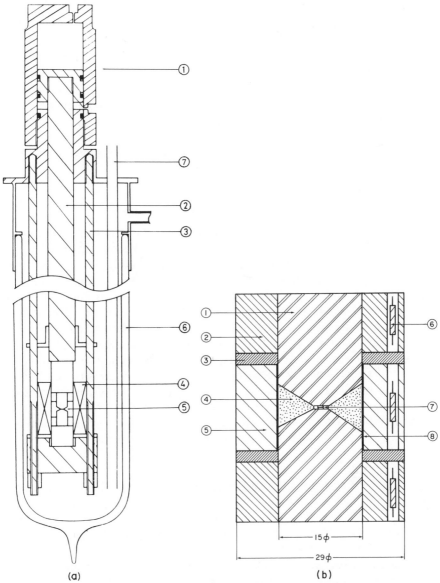

FIG. 4. Schematic diagram for measuring resistivity in the supported taper cell at cryogenic temperature. (a) Cryostat: (1) press head, (2) compression member, (3) tension member, (4) superconducting magnet, (5) supported taper cell, (6) helium dewar, and (7) helium siphon. (b) Cell: (1) sintered tungsten carbide piston, (2) stainless steel jacket, (3) Teflon, (4) pyrophyllite, (5) Be–Cu cylinder, (6) carbon resistor, (7) experimental sample, (8) Mylar.

The variations of electrical resistivity with pressure and temperature have been measured by the four-probe method with a supported taper anvil cell (Shimomura *et al.*, 1974; Minomura, 1981) and a diamond-anvil cell driven by helium gas pressure (Sakai *et al.*, 1982). A schematic diagram for measuring the electrical resistivity and superconductivity in the supported taper cell at cryogenic temperature is shown in Fig. 4. This cell consists of opposed anvils of sintered tungsten carbide or diamond and a cylinder of beryllium – copper alloy. The experimental sample and a pressure marker of bismuth or lead are compressed in silver chloride medium between the opposed anvils. This device is capable of generating pressures up to 500 kbar at low temperature.

III. Disorder-Induced Vibrational Spectra

Raman and infrared spectroscopy have proved to be powerful tools for understanding the atomic vibration in materials. Further discussion of Raman spectroscopy will be found in Volume 21B of this treatise. In diamond and zinc-blende structures there are three optical and three acoustic vibrational modes. The optical mode at the zone center is Raman-active but infrared-inactive. In amorphous materials, however, the momentum conservation rule is relaxed due to the structural disorder and all modes of vibration are Raman and infrared active. The first-order Raman and infrared spectra in a-Si and a-Si:H are explained by the crystalline vibrational density of states, which is appropriately broadened by the structural disorder. These spectra show the systematic changes in peak position, integrated intensity, and full width at half maximum (FWHM) with C_H or pressure (Shen *et al.*, 1980; Ishidate *et al.*, 1982). The experimental results are discussed in terms of changes in structural disorder as well as force constants.

The first-order Raman spectra of a-Si:H films prepared by glow-discharge decomposition of silane (GD) and reactive sputtering of c-Si target (SP) are shown in Fig. 5. The spectra are normalized to the same upper band. These spectra are interpreted as TO, LO, LA, and TA modes from the high-frequency side. One of the most interesting features is the decrease in intensity of the TA, LA, and LO bands with increasing C_H. The TO band shows the decrease in FWHM from 97 to 80 cm^{-1}. The change in Raman spectra with C_H is more significant in the GD films than in the SP films.

The pressure-induced effect on the first-order Raman spectra of SP a-Si:H film is shown in Fig. 6. The TO and LO bands shift to higher frequencies with compression at 25 kbar while the TA band shifts to lower frequencies. The LA band remains unchanged. The FWHM of the TO band decreases from 80 to 60 cm^{-1} with the compression whereas that of the TA

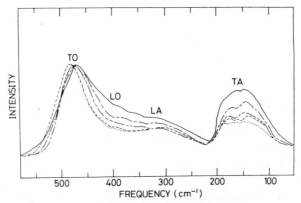

FIG. 5. Raman spectra of a-Si : H films prepared by glow-discharge decompositon (GD) and reactive sputtering (SP). —, SP a-Si pure; —·—, SP a-Si:H at 6 at. %; —··—, SP a-S:H at 17 at. %; ---, GD a-Si:H at 7 at. %; · · ·, GD a-Si:H at 18 at. %.

band increases from 90 to 100 cm^{-1}. As a consequence, the TA band shows an increase in integrated intensity in comparison with the TO band. This feature is in contrast to the alloying effect of H atoms where the intensity of the TA band decreases with increasing C_H. The pressure dependences of peak positions of the Raman spectra for a-Si : H films are listed in Table I.

The pressure-induced shift of the first- and second-order Raman spectra of c-Si has been measured by Weinstein and Piermarini (1975). The pressure dependences of Raman-active phonon frequencies for c-Si and a-Si : H are illustrated in Fig. 7. The pressure-induced effect on the peak position and FWHM of the TO and TA band of a-Si : H is much larger than that of the

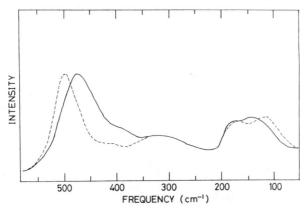

FIG. 6. Pressure effect on the Raman spectra of SP a-Si:H. SP a-Si:H at 24 at. %. —, 1 bar; ---, 25 kbar.

TABLE I

PEAK POSITIONS OF RAMAN SPECTRA FOR a-Si : H FILMS AT 1 BAR AND
THEIR PRESSURE DEPENDENCES

	TO	LO	LA	TA
Peak positions at 1 bar (cm⁻¹)	457 ± 5	380 ± 10	310 ± 5	150 ± 5
$(d\omega/dp)$ (cm⁻¹ kbar⁻¹)	1.0 ± 0.3	1.4 ± 0.6	0	-0.5 ± 0.3

crystalline vibrational density of states, which can be estimated from the
pressure dependences of Raman peaks at various symmetry points.

The Raman and infrared spectra of a-Si and a-Si : H relate to the crystalline vibrational density of states, which is appropriately broadened by the
structural disorder. The phonon dispersion curves of c-Si were determined
by inelastic neutron scattering (Dolling and Cowley, 1966). Alben *et al.*
(1975) calculated the vibrational density of states and the Raman and
infrared spectra for model structures of a-Si using a valence-force potential
function of the Keating form (1966):

$$V = \frac{3}{16}\frac{\alpha}{d^2}\sum_{bonds}\left(r_i^2 - d^2\right)^2 + \frac{3}{8}\frac{\beta}{d^2}\sum_{bond\ pairs}\left(\mathbf{r}_i \cdot \mathbf{r}_j + \frac{d^2}{3}\right)^2, \quad (1)$$

where α and β are the bond-stretching and bond-bending force constants,
respectively, and d is the strain-free equilibrium bond length. The Raman

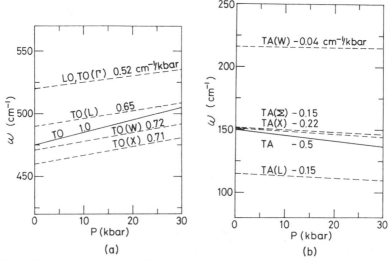

FIG. 7. Pressure dependences of Raman-active phonon frequencies for c-Si (dashed lines)
and a-Si : H (solid lines): (a) optical modes, (b) acoustic modes.

spectra of a-Si:H films under pressure show changes in peak position and FWHM of the TO and TA bands. The peak shift is associated with an increase in α and a decrease in β, and the change in FWHM is caused by the change in structural disorder. On the other hand, the alloying effect of H atoms shows a decrease in FWHM of the TO band and a decrease in intensity of the TA band. Since H atoms can be singly coordinated, they tend to relieve the mechanical strains due to cross-linking of the tetrahedrally bonded random Si network. These features may be interpreted as indicating a structural relaxation in bond length and bond angle. However, the pressure-induced effect may be interpreted as indicating a decrease in bond-length deviation and an increase in angular distortion.

IV. X-Ray Diffraction

The x-ray diffraction studies have been widely performed to obtain information about the local and intermediate-range order of amorphous materials. The x-ray diffraction intensity $I(k)$ of binary alloys is given by the form

$$I(k) = x_1 b_1^2 S_{11}(k) + 2\sqrt{x_1 x_2}\, b_1 b_2 S_{12}(k) + x_2 b_2^2 S_{22}(k), \qquad (2)$$

where $S_{ij}(k)$ is the partial interference function, b_i is the atomic scattering factor, x_i is the atomic fraction of i component and k is the wave vector. The $S_{ij}(k)$ can be determined from three independent measurements. The partial radial distribution function $g_{ij}(r)$ is derived from the Fourier transformation with respect to $S_{ij}(k)$. Actually, the first peak of $g_{ij}(r)$ could not be resolved completely unless $I(k)$ is measured at a wave vector on the order of 25 Å^{-1} or greater.

The x-ray diffraction studies of the atomic structure of a-Si:H have been reported by Dixmier et al. (1981) and Tsuji and Minomura (1981). The x-ray diffraction intensity profiles for a-Si and a-Si:H are shown in Fig. 8. Since the atomic scattering factor of H atoms is much smaller than that of Si atoms, the scattering intensity from H atoms is negligibly small. With increasing C_H the first and second diffraction peaks shift to lower angles and the intensity ratio of the first to second peak increases. The changes in peak position and intensity ratio with C_H are shown in Figs. 9 and 10. These features are interpreted as indicating a structural modification in the intermediate-range order.

The alloying effect of H atoms on the density d, first coordination number CN1, first coordination distance R_1, and optical energy gap E_o for the GD and SP films is illustrated in Fig. 11. The changes in d and CN1 with C_H for the GD films are smaller than those for the SP films. However, the changes in E_o with C_H for the GD films are larger than those for the SP films. In both alloys R_1 remains unchanged.

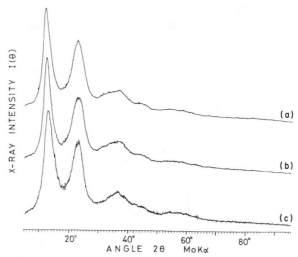

FIG. 8. X-ray diffraction intensity for (a) SP a-Si, (b) GD a-Si: H with 9 at. % H, and (c) GD a-Si: H with 21 at. % H.

The average values of CN1 for a-Si:H were calculated by Tsuji and Minomura (1981) from the following relations:

$$\begin{aligned}
CN1 &= (4c_0 + 3c_1 + 2c_2)/(c_0 + c_1 + c_2), \\
1 - C_H &= c_0 + c_1 + c_2, \\
C_H &= c_1 + 2c_2, \\
CN1 &= 4 - C_H/(1 - C_H),
\end{aligned} \tag{3}$$

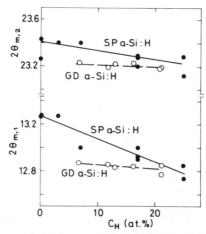

FIG. 9. Shift with pressure of the first and second diffraction peaks for GD a-Si: H (○) and SP a-Si: H (●).

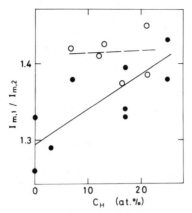

FIG. 10. Variation with H concentration (C_H) of the intensity ratio of the first to second diffraction peak for GD a-Si:H (O) and SP a-Si:H (●).

where CN1 is the Si concentration that is bonded to i H atoms. The calculated value of CN1 is 3.75 for 20 at. % H, which is in good agreement with the observed value for the GD films. However, the observed value decreases anomalously to 3.3 ± 0.1. Kubota *et al.* (1980) and Tanaka *et al.* (1980) have detected 6–8 at. % of Ar atoms for the SP films. Since the atomic scattering factor of Ar atoms lies close to that of Si atoms, the contribution of Ar atoms to CN1 must be taken into account. The anoma-

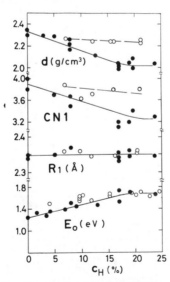

FIG. 11. Variation with C_H of the density d, first coordination number CN1, first coordination distance R_1, and optical energy gap E_o for GD a-Si:H (O) and SP a-Si:H (●).

lous decrease in CN1 for the SP films may be explained by the presence of Ar atoms, which contribute to zero CN1.

Postol *et al.* (1980) measured the neutron diffraction intensity for a-Si, a-Si:H, and a-Si:D prepared by magnetron sputtering to determine the $S_{ij}(k)$. The structural models for a-Si:H were constructed by adding H atoms randomly into the Si random network (Weaire *et al.*, 1979; Weaire and Wooten, 1980; Guttman, 1981). These models involve the calculated $S_{ij}(k)$ and $g_{ij}(k)$, which are in good agreement with the diffraction data. Weaire *et al.* (1979) suggested that H atoms are grouped in clusters of three or four atoms that are associated with small voids. Guttman (1981) showed that the average bond-length deviation decreases with decreasing β/α while the average angular distortion increases.

V. Phase Transitions

The primary effect of pressure is a decrease in interatomic distance. As a consequence, the application of pressure gives rise to a change in gap states and eventually to a phase transition from semiconductor to metal. The pressure-induced phase transitions in tetrahedrally bonded amorphous semiconductors have been studied by Shimomura *et al.* (1974), Minomura *et al.* (1977, 1980), and Minomura (1978, 1981). The phase transitions are demonstrated by discontinuous changes in optical energy gap, electrical resistivity, and atomic structure. After compression, the high-pressure

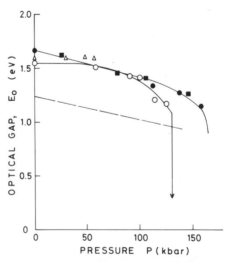

FIG. 12. Variation of the optical energy gap with pressure for GD and SP a-Si:H. (○, △) GD a-Si:H; (●, ■) SP a-Si:H; (— —) c-Si.

phases transform to various metastable modifications that consist of distorted tetrahedral bondings. These modifications have been used to construct microcrystalline models of amorphous materials (Joannopoulos and Cohen, 1973, 1976).

The optical absorption edge of a-Si:H films shifts to lower photon energies with increasing pressure. The optical energy gap E_0, which was determined from the extrapolation of $(\alpha h v))^{1/2}$ versus hv curves is plotted as a function of pressure in Fig. 12. The initial pressure dependence of E_0 is of the order of -1×10^{-3} eV kbar^{-1}, which is approximately equal to that of c-Si. The E_0 decreases more rapidly at pressures above 50 kbar and at a critical pressure of 130–150 kbar drops to zero. The GD films become opaque abruptly at about 130 kbar whereas the SP films become gradually opaque at about 150 kbar. After the compression, a large hysteresis is observed.

The variation of electrical resistivity with pressure for c-Si, a-Si, and a-Si:H is shown in Fig. 13. The phase transitions accompanied by a sharp drop in resistivity occur in c-Si at 150 kbar, in evaporated a-Si at 100 kbar, and in GD a-Si:H at 130 kbar. However, the SP a-Si:H shows the phase transitions accompanied by a sluggish decrease in resistivity at about 150 kbar. The pressure-induced phase transitions in the SP films are interpreted as indicating defects associated with impurities of H and Ar atoms.

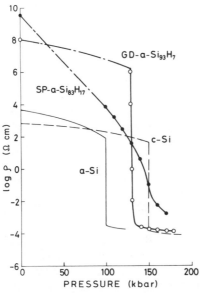

FIG. 13. Variation of the resistivity with pressure for c-Si, a-Si, and a-Si:H.

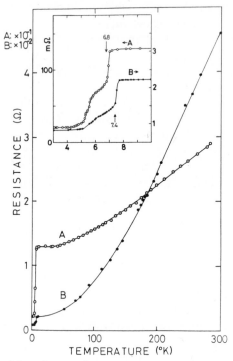

FIG. 14. Variation of the resistance with temperature for the high-pressure phases of (A) GD a-Si:H and (B) c-Si at 170 kbar.

The metallic character of the high-pressure phases of crystalline and amorphous Si and Ge has been demonstrated as superconductivity (Wittig, 1966; Shimomura *et al.*, 1974; Sakai *et al.*, 1982). The variation of resistance with temperature for the high-pressure phases of c-Si and GD a-Si:H at 170 kbar is shown in Fig. 14. They become superconducting with a sharp drop in resistance at transition temperatures of 6.8 and 7.4°K, respectively. The superconducting transition temperatures of crystalline and amorphous group IV elements are plotted as a function of pressure in Fig. 15. The transition temperatures decrease with increasing pressure. The observed values for the high-pressure phases of c-Si, a-Si, and a-Si:H do not change appreciably.

The variation of x-ray diffraction intensity with pressure for evaporated a-Si is shown in Fig. 16. The first and second diffuse peaks shift to higher angles with increasing pressure and broaden appreciably. At about 100 kbar new crystalline-like peaks appear over the amorphous background. They grow and develop a finer structure with increasing pressure. The first and second crystalline-like peaks at pressures below 150 kbar correspond to the

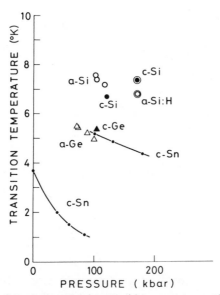

FIG. 15. Variation of the superconducting transition temperature with pressure for crystalline and amorphous group-IV elements. c-Sn transforms from a β-Sn structure to a simple body-centered tetragonal structure at about 100 kbar.

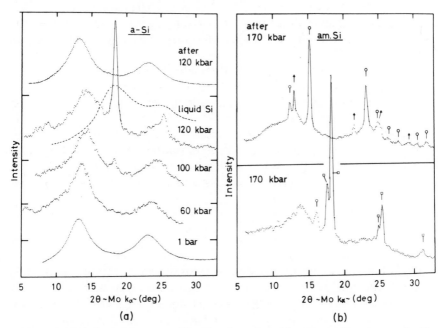

FIG. 16. Variation of the x-ray diffraction intensity with pressure for evaporated a-Si. (a) 120 kbar and below; (b) 170 kbar and above. $\square$, β-Sn; $\bigcirc$, BC-8; $\blacktriangle$, diamond.

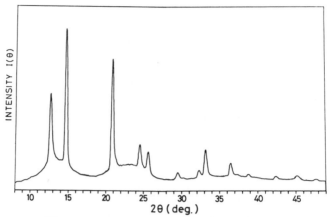

FIG. 17. X-ray diffraction pattern for the high-pressure modification of SP a-Si : H after compression at 180 kbar.

(101) and (211) spacings of β-Sn structure, respectively. On releasing the pressure, these modifications transform to the tetrahedrally bonded amorphous state. On the other hand, the high-pressure phases above 150 kbar transform to a body-centered cubic structure (BC-8).

The high-pressure phases of a-Si : H below 150 kbar can transform to the semiconducting amorphous state on releasing the pressure. However, the high-pressure phases above 150 kbar transform to a tetragonal structure, as shown in Fig. 17.

Jamieson (1963) reported that c-Si under pressure shows a structural phase transition to a β-Sn type with lattice constants: $a = 4.684$ Å, $c = 2.585$ Å, and $c/a = 0.5516$. In this structure, each atom has four nearest neighbors at 2.430 Å and two next nearest neighbors at 2.585 Å. The change of structure can be accomplished by displacing atoms along the c axis with an increase in bond length and a decrease in volume. On the other hand, a-Si and a-Si : H under pressure show a structural phase transition to a distorted β-Sn type. The change of structure is obviously heterogeneous with mixing of low- and high-pressure phases.

VI. Conclusion

Pressure-induced effects and phase transitions in plasma-deposited a-Si : H films have been investigated by measurements of Raman scattering, x-ray diffraction, fundamental optical absorption, electrical resistivity, and superconductivity. The alloying effect of H atoms on the Raman spectra and x-ray diffraction intensity profiles have been also studied by comparing the pressure effects. The changes in Raman spectra with hydrogen content and pressure are interpreted as indicating at least two independent terms of structural disorder. It is proposed that the incorporation of H atoms gives rise to a structural relaxation in terms of bond-length and bond-angle distortion while the application of pressure causes a decrease in bond-length deviation and an increase in angular distortion.

The hydrogen environment and defects in a-Si : H films are reflected in the pressure-induced phase transitions and the associated changes in optical gap and electrical resistivity. It is shown from x-ray diffraction experiments that the phase transitions are accompanied by a change of structure to that of a distorted β-Sn type.

The high-pressure studies have improved our understanding of the local order of a-Si : H films. However, in the structural studies it is necessary to develop a variety of high-pressure experiments. For example, in the future the following studies should be carried out. First, the compressibility has to be determined by measurements of the Brillouin scattering. Second, the impurity environment should be studied by measurements of EXAFS and NMR. Third, the gap states should be studied by measurements of the time-resolved luminescence and reflectance.

REFERENCES

Alben, R., Weaire, D., Smith, J., and Brodsky, M. H. (1975). *Phys. Rev. B* **11**, 2271–2296.
Brodsky, M. H., Cardona, M., and Cuomo, J. J. (1977). *Phys. Rev. B* **16**, 3556–3571.
Dixmier, J., Mosseri, R., and Sadoc, J. F. (1981). *J. Phys. (Orsay, Fr.)* **42**, C4-237–C4-240.
Dolling, G., and Cowley, R. A. (1966). *Proc. Phys. Soc., London* **88**, 463–494.
Fujii, Y., Shimomura, O., Takemura, K., Hoshino, S., and Minomura, S. (1980). *J. Appl. Crystallogr.* **13**, 284–289.
Guttman, L. (1981). *Phys. Rev. B* **23**, 1866–1874.
Ishidate, T., Inoue, K., Tsuji, K., and Minomura, S. (1982). *Solid State Commun.* **42**, 197–200.
Jamieson, L. C. (1963). *Science* **139**, 762–764.
Joannopoulos, J. D., and Cohen, M. L. (1973). *Phys. Rev. B* **7**, 2644–2657.
Joannopoulos, J. D., and Cohen, M. L. (1976). *Solid State Phys.* **31**, 71–148.
Keating, P. N. (1966). *Phys. Rev.* **145**, 637–645.
Kubota, K., Imura, T., Iwami, M., Hiraki, H., Satou, M., Fujimoto, F., Hamakawa, Y., Minomura, S., and Tanaka, K. (1980). *Nucl. Instrum. Methods* **168**, 211–215.
Lucovsky, G., Nemanich, R. J., and Knights, J. C. (1979). *Phys. Rev. B* **19**, 2064–2073.
Matsuda, A., and Tanaka, K. (1982). *Thin Solid Films* **92**, 171–187.
Minomura, S. (1978). *Symp. High-Pressure Low-Temp. Phys. 1977* pp. 483–503.
Minomura, S. (1981). *J. Phys. (Orsay, Fr.)* **42**, C4 181–C4188.
Minomura, S., Shimomura, O., Asaumi, K., Oyanagi, H., and Takemura, K. (1977). *Amorphous Liq. Semicond. Proc. Int. Conf. 7th, 1977* pp. 53–57.
Minomura, S., Tsuji, K., Oyanagi, H., and Fujii, Y. (1980). *J. Non-Cryst. Solids* **35–36**, 513–518.
Piermarini, G. J., Block, S., Barnett, J. D., and Forman, R. A. (1975). *J. Appl. Phys.* **46**, 2774–2780.

Postol, T. A., Falco, C. M., Kampwirth, R. T., Schuller, I. K., and Yelon, W. B. (1980). *Phys. Rev. Lett.* **45**, 648–652.

Sakai, N., Kajiwara, T., Tsuji, K., and Minomura, S. (1982). *Rev. Sci. Instrum.* **53**, 499–502.

Shen, S. C., Fang, C. J., Cardona, M., and Genzel, L. (1980). *Phys. Rev. B Condens. Matter* [3] **23**, 2913–2919.

Shimomura, O., Minomura, S., Sakai, N., Asaumi, K., Tamura, K., Fukushima, J., and Endo, H. (1974). *Philos. Mag.* [8] **29**, 547–558.

Takemura, K., Shimomura, O., Tsuji, K., and Minomura, S. (1979). *High Temp.–High Press.* **11**, 311–316.

Tanaka, K., Yamazaki, S., Nakagawa, K., Matsuda, A., Okushi, H., Mastumura, M., and Iizima, S. (1980). *J. Non-Cryst. Solids* **35–36**, 475–480.

Tsuji, K., and Minomura, S. (1981). *J. Physique* **42**, C-4 233–236.

Weaire, D., and Wooten, F. (1980). *J. Non-Cryst. Solids* **35–36**, 495–500.

Weaire, D., Higgins, N., Moore, P., and Marshall, I. (1979). *Philos. Mag. [Part] B* **40**, 243–245.

Weinstein, B. A., and Piermarini, G. J. (1975). *Phys. Rev. B* **12**, 1172–1189.

Wittig, J. (1966). *Z. Phys.* **195**, 215–227.

CHAPTER 14

Defects and Density of Localized States

David Adler

DEPARTMENT OF ELECTRICAL ENGINEERING AND COMPUTER SCIENCE
AND CENTER FOR MATERIALS SCIENCE AND ENGINEERING
MASSACHUSETTS INSTITUTE OF TECHNOLOGY
CAMBRIDGE, MASSACHUSETTS

I. Normal Structural Bonding

1. STRUCTURE

In general, amorphous solids can be considered simply as giant chemical molecules that are capable of filling space. As is true in all molecules, the physical properties are predominantly controlled by the chemical nature of the constituent atoms. However, when $\sim 10^{24}$ atoms are present, there are clearly enormous numbers of possibilities for isomers with sharply different electronic behavior. In principle, a complete approach to the problem would entail (1) a determination of the *structure* of the solid, viz. the equilibrium positions of the atoms; (2) the evaluation of the normal modes of vibration around these equilibrium positions, that is, the *phonon* structure; and (3) the deduction of the excited *electronic structure,* ordinarily approximated by an effective one-electron density-of-states for the solid, $g(E)$. In all solids the structure is always determined empirically. In carefully grown crystals the near-perfect periodicity provides major simplifications in both the experimental interpretation and the theoretical analysis, and often a detailed understanding of the physical properties is possible. In contrast, amorphous solids do not exhibit sufficient periodicity to provide any significant theoretical or experimental simplifications. Furthermore, they are necessarily prepared by nonequilibrium methods to avoid obtaining the crystalline phase (which generally exhibits the lowest free energy at the temperatures and pressures employed), and thus their structure can be very sensitive to the details of the preparation conditions. Since the electronic density of states depends strongly on the precise atomic structure, the complexity of the electronic problem can be compounded by the study of materials prepared in different manners. It should also be borne in mind that small atomic motions between local equilibrium positions can be induced by perturbing the material; this can be accomplished, for example, thermally, by absorption of light, or by injection of excess carriers.

2. TOPOLOGY

Since the chemical bonding in hydrogenated amorphous silicon (a-Si:H) is predominantly *covalent,* there are strong forces locally constraining (1) the number of nearest neighbors Z, (2) the bond lengths (Si–Si and Si–H), and (3) the bond angles (Si only). Normal structural bonding is said to occur whenever Z is optimized for the constituent atom under consideration. In pure a-Si:H, the optimum value of Z is unity for H and four for Si; the average coordination number for any particular alloy a-Si$_{1-x}$H$_x$ is thus optimally

$$\overline{Z} = 4 - 3x. \tag{1}$$

In general, since each covalent bond forms between two atoms, a fixed bond length places an average of $\overline{Z}/2$ constraints on the atoms in an amorphous network. Since each atom has three spatial degrees of freedom, the bond lengths can all be optimized provided

$$\overline{Z} \leq 6.$$

This is clearly the case in all a-Si:H alloys. However, there are also somewhat weaker chemical forces constraining the bond angles. Since the average number of bond angles per atom is $\overline{Z}(\overline{Z} - 1)/2$, the total number of constraints per atom if both bond lengths and bond angles are to be optimized is approximately $\overline{Z}^2/2$. Thus the constraints can all be satisfied only if (Phillips, 1979a)

$$\overline{Z} < \sqrt{6} = 2.45.$$

This is certainly not the case for a-Si:H alloys of interest, and we can conclude that a-Si:H films all exhibit a considerable amount of strain. In most cases these strains will be relieved by bond-angle distortions. Indeed, there is structural evidence in a-Si films (Moss and Gracyzk, 1970) for deviations of $\pm 10°$ from the optimal $109.5°$ bond angles appropriate to sp³ bonding. If the average coordination number were lower than 2.45, we might even expect some dihedral angle order (Adler, 1982a). Such order has been found in a-Se ($Z = 2$), but should not be expected in a-Si:H. In fact, there is very little structural evidence for well-defined third-neighbor distances in a-Si films (Moss and Graczyk, 1970), suggesting near-complete dihedral-angle disorder.

3. BAND STRUCTURE

Since the vast majority of atoms in a-Si:H exhibit normal structural bonding, the essential features of the band structure of these alloys can be estimated from the more easily calculated states of crystalline Si (c-Si). However, since H concentrations of 10–30% are typical, it is dangerous to neglect the effects of the hydrogen. In many solids a tight-binding calculation often provides an initial qualitative understanding to the basic band structure (Adler, 1980a). For a Si atom surrounded by four Si atoms in a tetrahedral (sp³) configuration, the tight-binding approach is sketched in Fig. 1a. The eight hybridized sp³ orbitals are split by the four covalent Si–Si bonds into four lower bonding orbitals and four higher antibonding orbitals. The four outer electrons on each Si atom are sufficient to fill the bonding orbitals. In the solid, second and farther neighbor interactions spread the bonding orbitals into the (filled) valence band and the antibonding orbitals into the (empty) conduction band. When a Si atom is surrounded by three Si atoms and one H atom in a tetrahedral configuration, the situation is as

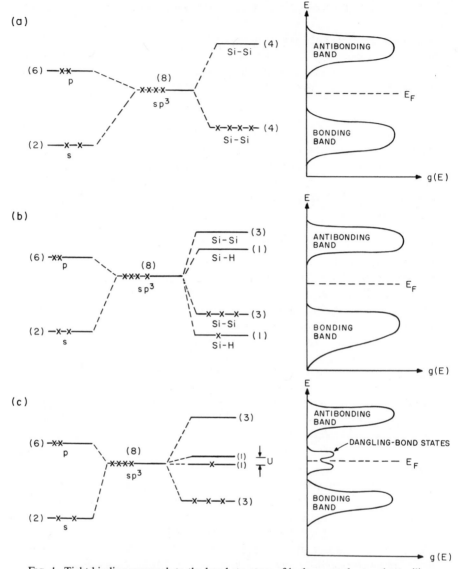

FIG. 1. Tight-binding approach to the band structure of hydrogenated amorphous silicon films. (a) A central Si atom tetrahedrally coordinated by four neighboring Si atoms; the eight sp³ hybridized orbitals are split into four bonding and four antibonding orbitals, which respectively spread into the valence and conduction bands of the solid. (b) A central Si atom tetrahedrally coordinated by three neighboring Si atoms and one H atom; the Si–H bonding orbital falls deep within the valence band, while the Si–H antibonding orbital lies near the lower edge of the conduction band. (c) A central Si atom surrounded by only three neighboring Si atoms forming hybridized sp³ bonds. The fourth electron on the central atom occupies a nonbonding orbital (a dangling bond), separated from its unoccupied partner by the correlation energy U; both nonbonding states lie within the gap.

sketched in Fig. 1b. The Si–H bond strength (3.4 eV) is greater than the Si–Si bond strength (2.4 eV), resulting in a greater bonding–antibonding splitting from the former. However, H is more electronegative than Si, so that the 1s orbital of H lies below the 3sp³ hybridized orbital of Si, suggesting that the presence of H should primarily induce a sharp reduction in the energy of some states near the top of the valence band. This shift of weight in the density of states should serve to increase the energy gap with increasing H concentration. Indeed, this is in agreement with experimental results (Cody et al., 1980); the optical gap increases from about 1.5 eV in pure a-Si to about 2.0 eV when 30% H is present. There is both experimental (von Roedern et al., 1977, 1980) and theoretical (Joannopoulos, 1980; Allen and Joannopoulos, 1980; Johnson et al., 1980; Papconstantopoulos and Econo-mou, 1981) evidence that the origin of this increase in gap is a reduction in the energy of the states near the top of the valence band. There are also some experimental (Moustakas et al., 1977) and theoretical (Ching et al., 1979, 1980) indications that the Si–H antibonding orbitals are energetically near the bottom of the conduction band in a-Si:H alloys.

4. BAND TAILS

Theoretically, the lack of long-range order inherent in any amorphous solid must result in the existence of valence and conduction *band tails*, even if only normal structural bonding exists (Adler, 1980b). From a chemical point of view, these band tails can arise from bond-length variations, bond-angle distortions, or dihedral-angle disorder. Stretched Si–Si bonds could contribute to the band tails, but these could be passivated by the formation of three-center bonds with hydrogen and thereby removed from the tails (Eberhart et al., 1982). There is some theoretical evidence that bond-angle distortions contribute to the band tails by shifting states from the bulk of the bands to the regions near the edges (Joannopoulos, 1980). Alternatively, virtual-crystal-approximation estimates (Singh, 1981) have indicated that dihedral-angle disorder can be the origin of significant band tailing. In addition, some of the defect configurations that will be discussed throughout the remainder of the chapter could also contribute to the band tails.

II. Defects in Solids

5. THERMODYNAMICALLY INDUCED DEFECTS

Solids are ordinarily prepared under conditions for which thermo-dynamic variables such as the pressure P, the temperature T, and the chemical potential of atomic or molecular species i, μ_i, can be defined.

When solids are obtained by cooling slowly from the liquid phase, long-range atomic motions are frozen in at the melting temperature T_m. For simple elemental materials and binary alloys, this always yields a periodic array of atoms, that is, a *crystal*. For crystals, not only is the local environment of any given atom known, but once the nature of the atom and the orientation of its nearest neighbors is determined, the positions of all the distant atoms are known. Nevertheless, even at best, the order is far from perfect. Point defects, line defects, and defect clusters always exist in some concentrations. This is particularly true if a given defect has a relatively low creation energy, ΔE, since the concentration of any given defect frozen in at the melting point is given by the thermodynamic relationship:

$$N_d = N_0 \exp(-\Delta E/kT_m), \tag{2}$$

where N_0 is the total concentration in the solid. For example, if $\Delta E = 1$ eV, $T_m = 1000°$K, and $N_0 = 10^{23}$ cm^{-3}, $N_d \sim 10^{18}$ cm^{-3}.

Amorphous solids can often be quenched from the liquid by very rapid cooling. Some supercooling then takes place, and the viscosity does not sharply increase until the glass transition temperature T_g is reached. Below T_g, the amorphous solid thus formed is called a *glass*. The concentration of thermodynamically induced defects is then given by a relationship analogous to Eq. (2) with T_g replacing T_m. Since $T_g < T_m$, the thermodynamically induced defect concentrations are generally lower in glasses than in crystals.

Hydrogenated amorphous silicon is neither a crystal nor a glass and is always prepared by direct deposition from the vapor onto a cold substrate. The only characteristic temperature that can be defined during the deposition process is the substrate temperature T_s. We can still define a concentration of thermodynamically induced defects, given by

$$N_d = N_0 \exp(-\Delta E/kT_s). \tag{3}$$

However, we would expect that N_d calculated from Eq. (3) just represents a lower limit to the actual defect concentration in vapor-deposited amorphous solids.

6. STRAIN-INDUCED DEFECTS

In addition to those required by thermodynamics, defects can be introduced into any solid by strains during the deposition or growth process. For amorphous materials with $\bar{Z} < 2.4$, such strains can be minimal if care is taken in their preparation. However, if $\bar{Z} > 3$, as is the case for a-Si:H, a great deal of strain is necessarily present, and we might expect the incorporations of defects that tend to lower the average coordination number.

As discussed in Section 5, the thermodynamically induced defects obey Eq. (3), and are thus predominated by the particular defect with the lowest creation energy, ΔE. In the unlikely event that two distinct types of defects have very similar values of ΔE, they both will be present with similar concentrations, but this would not be expected to occur very often. In contrast, many different types of strain-induced defects can exist simultaneously in films with $\bar{Z} > 3$. Consequently, there can be large differences in the physical properties of materials prepared under different deposition conditions, and it can be very difficult to pinpoint the nature of the defects that are present.

7. ELECTRONIC STRUCTURE OF DEFECTS

Since the vast majority of atoms in any solid, crystalline or amorphous, participate in normal structural bonding, the density of electronic states originating from such atoms is ordinarily relatively large, $g(E) > 10^{21}$ cm^{-3} eV^{-1}. Grossly distorted bond angles could contribute to band-tail states with lower values for $g(E)$. It is likely that a critical value of $g(E)$ exists below which the states can be considered to be essentially localized (Mott and Davis, 1979; Adler, 1982b); these critical energies are called *mobility edges.*

In contrast, states originating from atoms in defect configurations generally exhibit much lower values of $g(E)$. When these fall in regions of energy that do not overlap those within the mobility edges of a band of extended states, the resulting states are *localized;* this is true whether the solid is crystalline or amorphous. Localized defect states that lie within the *mobility gap,* that is, that have energies above the upper mobility edge of the valence band E_v and the lower mobility edge of the conduction band E_c, ordinarily control the transport properties of any given semiconductor. Such states determine the position of the Fermi energy E_F, which controls the free-carrier concentrations at thermal equilibrium, and they also act as the traps and recombination centers that determine the kinetics for restoration of equilibrium after any perturbations in these concentrations (Rose, 1978).

A major difficulty arising from the presence of localized states within the mobility gap is the necessary breakdown of the two major approximations of the band theory of solids: the one-electron approximation and the adiabatic approximation (Adler, 1982c). The one-electron approximation neglects the possibility that two electrons can correlate their motion to minimize their mutual electrostatic repulsion. The repulsion between two electrons with opposite spins that are simultaneously present in the same spatial state is usually called the *correlation energy U.* If the state is extended, U is ordinarily quite small and can be neglected. However, if the state is localized, U can be in the 0.1–1-eV range and it is then of major significance. When correlations are important, the density of states necessarily depends on the state of occupation: if an electron of either spin is present, the corresponding state for the electron with opposite spin is increased in energy by U (Adler, 1982d).

The adiabatic approximation neglects electron–phonon interactions, which are ordinarily small for carriers in extended states in nonpolar solids. However, the presence of an excess electron or hole localized near an atom effectively changes the nature of that atom to one in a different column in the Periodic Table, at least as far as its optimal chemical bonding is concerned. This could induce relaxations of the surrounding atoms, resulting in shifts in the energies of the localized states. Such relaxations can, in turn, introduce other complications. Electronic effects take place at typical frequencies of 10^{15} Hz, so that variations in the effective density of states $g(E)$ due to correlations can be thought of as essentially instantaneous. However, typical phonon frequencies are of the order of 10^{12} Hz, very slow compared to optical frequencies. This results in a different $g(E)$ obtained from optical experiments compared to that obtained from electrical measurements. Furthermore, the local relaxations induced by changes in electronic occupancy can be retarded by the presence of significant potential barriers, which serve to greatly slow the effects of electron–phonon induced energy shifts. Such barriers can yield complex kinetics that are extremely temperature dependent.

The possibility also exists that free carriers in the conduction or valence bands can induce significant atomic relaxations due to the electron–phonon coupling. If these relaxations are extended, they just result in relatively small changes in the electronic energies and effective masses. These changes can be taken into account simply by renormalizing the band parameters; the carriers are then said to form *large polarons* (Frölich, 1954). Alternatively, if the atomic relaxations do not extend very far spatially, the would-be free carrier can be localized by the motions induced by their own presence, yielding what has come to be known as a *small polaron* (Holstein, 1959). Small polarons in disordered solids conduct only by phonon-assisted hopping between localized states, transport necessarily characterized by relatively low values of carrier mobility ($\mu < 0.1$ cm^2 V^{-1} sec^{-1}).

III. Defects in Pure a-Si:H

8. DANGLING BONDS

It is clear that since a-Si:H forms an overconstrained network those defects that are introduced via strains during the film deposition process should tend to reduce the average coordination. In any event, since tetrahedral coordination is the maximum possible using s and p electrons only (and d-hybridization is not common for Si unless strongly electronegative atoms such as F are present in large concentrations), overcoordinated Si defects are not likely. The simplest defect that might be expected to be present is the isolated dangling bond. We use the notation (Adler, 1978) T_z^q, where T stands for a Si atom (a tathogen atom from column IV of the Periodic Table), the subscript Z signifies the coordination number, and the superscript q gives the local charge state. Thus normal structural bonding is represented as T_4^0 (see Fig. 2a) and a neutral dangling bond is T_3^0 (see Fig. 2b). Since a dangling bond is not the optimal state for a neutral Si atom, it is not clear how the bond lengths and bond angles relax in the T_3^0 configuration. There is some theoretical evidence (Joannopoulos, 1980) that the bonding remains sp³, thus retaining the tetrahedral bond angle of 109.5°.

Tight-binding estimates (Adler, 1978) suggest that the T_3^0 defect yields two states in the mobility gap, a lower filled state and an upper empty state (Fig. 1c). In the absence of significant atomic relaxations, these two states differ in

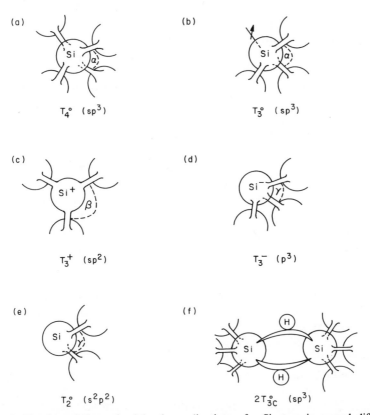

FIG. 2. Sketches of the optimal local coordinations of a Si atom in several different configurations; (a) ground state, T_4^0; (b) neutral dangling bond, T_3^0; (c) positively charged dangling bond, T_3^+; (d) negatively charged dangling bond, T_3^-; (e) two fold-coordinated Si atom, T_2^0; (f) complex consisting of two three-center bonds with bridging H atoms. The bond angles identified are $\alpha = 109.5°$, $\beta = 120°$, $\gamma \approx 95°$.

energy by the correlation energy U. However, if an electron is removed from a T_3^0 center, it converts to T_3^+ (Fig. 2c), which induces strong chemical forces that tend to distort the local environment. A positively charged Si ion is isoelectronic to Al, which optimally bonds sp^2, in a planar configuration with a 120° bond angle. This results in an *increase* in the energy of the now-empty lower dangling-bond orbital, which most likely has nearly pure p character after the distortion. This process thus represents an effective stabilization of the holes that are trapped by T_3^0 center; it converts to T_3^- (Fig. 2d). But a negatively charged Si ion is isoelectronic to P, which optimally forms predominately p bonds. The chemical forces thus act to decrease the bond angle to about 95°, by moving the Si$^-$ ion away from the plane of its three nearest neighbors. The resulting dehybridization tends to *lower* the energy of the now-filled upper dangling-bond orbital. Thus, local atomic relaxations result in an effective decrease of the correlation energy. This can be taken into account by renormalizing U to a lower value, U_{eff}. The possibility exists that U_{eff} can, in fact, be negative (Anderson, 1975). This occurs for the case of defects in chalcogenides such as As$_2$Se$_3$ (Kastner *et al.*, 1976), with major consequences as far as the transport behavior of the materials is concerned. The sign of U_{eff} for the dangling bond in a-Si:H is still a matter of controversy. Tight-binding estimates (Adler, 1978) suggest that U_{eff} could be negative, but more sophisticated calculations (Allen and Joannopoulos, 1980) indicate it remains positive. If $U_{eff} > 0$, the resulting electronic band structure for a-Si:H characterized by only dangling-bond defects would be as sketched in Fig. 3a; the lower, filled T_3^0 band would contribute an ESR signal from the unpaired spins on the dangling bonds. In contrast, if $U_{eff} < 0$, $g(E)$ would instead be as sketched in Fig. 4a. In this case, E_F would be pinned to the extent of the dangling-bond concentration (Adler and Yoffa, 1976), and the unpaired-spin concentration, N_s, should be quite small.

ESR data (Dersch *et al.*, 1981) clearly indicate the presence of a well-defined unpaired spin with $g = 2.0055$ and a peak-to-peak linewidth of 7.5 G, almost universally associated with T_3^0 centers. This is evidence for a positive correlation energy in a-Si:H. However, since a-Si:H is an overconstrained network and, in fact, these excess strains are the origin of the dangling bonds in the first place, another explanation is possible (Adler, 1983). If $U_{eff} < 0$ when complete atomic relaxations around both the T_3^+ and T_3^- centers can take place, but such relaxations (which require bond-angle changes from the optimal tetrahedral 109.5° angles to the vicinities of 120° and 95°, respectively) are retarded in particularly strained regions of the film, then isolated T_3^0 centers could coexist with T_3^+–T_3^- pairs. The T_3^0 centers then account for the ESR signal. There is some evidence that this model is valid in a-Si:H films, and it will be discussed further in Part VI.

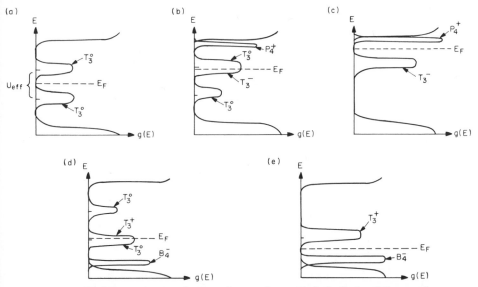

FIG. 3. Effective one-electron density of states for a tetrahedrally bonded amorphous semiconductor with dangling-bond defects characterized by $U_{eff} > 0$: (a) undoped sample, (b) moderately P-doped sample, (c) heavily P-doped sample, (d) moderately B-doped sample, (e) heavily B-doped sample. The defect bands are labeled by their state when *filled* if they are located *below* E_F and when *empty* if they are located *above* E_F.

9. TWOFOLD-COORDINATED SILICON CENTERS

The ground-state electronic configuration of the Si atom is s^2p^2 (see Fig. 1). The lowest-energy covalent-bonding configuration is a tetrahedral arrangement in which hybridized sp^3 orbitals are used to bond to four neighbors, that is, T_4^0. This costs the s–p promotion energy, but allows the formation of *two* extra bonds; in addition all four hybridized bonds are stronger than the p bonds that would be formed if no promotion occurred. However, when we are considering the relative creation energies of defect centers, we must compare the energy of the two fold-coordinated center, T_2^0 (see Fig. 2e), to that of the dangling bond, T_3^0. In this case it is not clear that the s–p promotion energy (about 6 eV for the isolated atom) is repaid by the formation of only one additional bond (and the strengthening of the other two bonds after the hybridization). In fact, chemically it seems unlikely that T_3^0 centers have a lower creation energy than T_2^0 centers (Adler, 1978). Divalent Si is common in many compounds, notably SiH_2, whereas trivalent neutral Si is less common. Furthermore, since T_2^0 centers lower the average coordination number twice as effectively as T_3^0 centers, they are more efficient in relieving strains in the film.

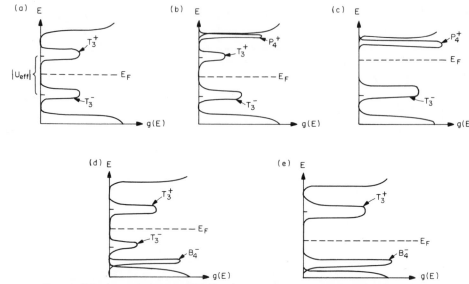

FIG. 4. Effective one-electron density of states for a tetrahedrally bonded amorphous semiconductor with well-separated dangling-bond defects characterized by $U_{eff} < 0$: (a) undoped sample, (b) moderately P-doped sample, (c) heavily P-doped sample; (d) moderately B-doped sample, (e) heavily B-doped sample. The notation is the same as in Fig. 1.

The electronic structure of the T_2^0 center is quite complex, since *five* different charge states are possible, T_2^{2+}, T_2^+, T_2^0, T_2^-, and T_2^{2-}. It is likely that all of these states fall within the mobility gap of a-Si:H. There is no question that the effective correlation energy is positive for T_2^0 defects, since the reaction,

$$2T_2^0 \rightarrow T_2^+ + T_2^-$$

is clearly endothermic (Adler, 1978). Since T_2^0 centers contain only paired spins, no ESR signal results from this charged state. Only T_2^+ and T_2^- centers contain unpaired spins.

If T_2 and T_3 centers are simultaneously present, complex charged centers can exist. Two possibilities are $T_2^+ - T_3^-$ pairs and $T_2^{2+} - 2T_3^-$ triplets. The former forms a dipole and the latter a quadrupole, and both should exhibit spatial correlations from the varying electrostatic forces. Clearly, the situation can be quite complicated.

10. THREE-CENTER BONDS

The most obvious effects of hydrogen incorporation in a-Si:H are the lowering of the average coordination number and the passivation of dangling bonds, as discussed previously. The formation of strong covalent Si–H bonds can convert both T_3^0 and T_2^0 centers to normal structural bonding, T_4^0. Even T_3^+–T_3^- pairs can be converted to $2T_4^0$ by bonding with hydrogen. The facts that hydrogenation of pure a-Si films requires about 100 times as much H as the unpaired-spin concentration suggests (Sol *et al.*, 1980) and that about 100 times as much H is given off on heating a-Si:H films than unpaired spins are created (Biegelsen *et al.*, 1979) strongly indicates that spinless defects are present in the alloys (Adler, 1981).

Ovshinsky and Adler (1978) suggested the presence of three-center bonds with bridging hydrogen atoms, as sketched in Fig. 2f. Recent calculations (Eberhart *et al.*, 1982) indicate that such bonds can remove any localized states resulting from stretched Si–Si bonds from the mobility gap. In this sense, three-center bonds should not be considered defect centers, but rather a type of normal structural bonding. Their importance may be more in understanding the apparent absence of stretched bonds in a-Si:H than in any effects on the electronic structure. In addition, such a center could well account for the observed H_2 rotational modes observed in NMR experiments (Carlos and Taylor, 1982).

11. INTIMATE PAIRS

When charged dipoles such as T_3^+–T_3^- or T_2^+–T_3^- pairs are present, it is clear that their creation energy can be lowered by reducing their spatial separation (Kastner *et al.*, 1976). The maximum reduction in ΔE occurs when the oppositely charged centers are nearest neighbors, in which case the defect complex is called an intimate charge-transfer defect or ICTD (Adler and Yoffa, 1977). In general, the creation energy of a pair separated by a distance R is reduced by a value of $e^2/\kappa R$, where κ is the effective dielectric constant. Thus an ICTD is stabilized by an energy $M = e^2/\kappa R_0$, where R_0 is the nearest-neighbor separation of 2.35 Å. If we use the full dielectric constant of a-Si:H, $\kappa \cong 12$, we can estimate $M \simeq 0.5$ eV; however, the real value is probably considerably larger, due to the effective reduction in screening at small distances.

When spatially correlated pairs are present, the electronic structure is considerably complicated because of changes in the electrostatic interaction upon variations in occupancy (Adler, 1980c). For an ICTD, the two levels associated with a distant pair (see Fig. 4a) are each split; there are a total of four levels, resulting in five possible conditions of occupancy (from T_3^+–T_3^+ to T_3^-–T_3^-). In this case, E_F unpins to the extent of $2M$ (Adler and Yoffa, 1977).

In general, the stabilization of oppositely charged defects depends on their separation, R. Although the law of mass action cannot be expected to apply when the defects are strain-induced as in a-Si:H, it does suggest that an approximately exponential distribution of separations should characterize

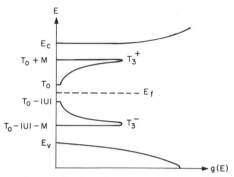

FIG. 5. Effective one-electron density of states for a tetrahedrally bonded amorphous semiconductor containing a distribution of spatially correlated dangling-bond defects with $V_{eff} < 0$. T_0 represents the energy of the nonbonded electron on a neutral dangling bond and M is the electrostatic attraction between a T_3^+ center and a T_3^- center located on nearest-neighboring sites.

such films (Adler, 1981). The effective density of states for undoped a-Si:H films characterized only by $T_3^+ - T_3^-$ defects is then shown in Fig. 5. Note that if M is actually as large as 0.9 eV or more, the defect levels cannot be distinguished from exponential band tails.

The possibility exists that $U_{eff} > 0$ for isolated dangling bonds, but $M > U_{eff}$. This leads to a stabilization of close $T_3^+ - T_3^-$ pairs as well as isolated T_3^0 centers (Elliot, 1978; Adler, 1981). Such a model provides another explanation of the ESR data discussed previously. In this case, Fig. 5 must be modified so that the two bands overlap near the center of the gap, the region of overlap representing the T_3^0 centers with $U_{eff} > 0$. For such a situation, E_F can be varied easily by the induction or injection of excess charge or by doping.

12. DEFECT COMPLEXES

More complicated defects than ICTDs can also be present in pure a-Si:H films. One possibility, a $T^{2+} - 2T_3^-$ quadrupole has already been mentioned; this defect can have *nine* different charge conditions, although it is highly unlikely that all eight defect levels would appear within the 1.8-eV gap of a-Si:H.

Vacancy-type defects have often been proposed for a-Si:H films (Hirose, 1981; Chakraverty, 1981), although it is not yet clear what their structure might be in an amorphous network. In high-quality a-Si:H films, it might be expected that four hydrogen atoms would compensate the potential dangling bonds in vacancy-type defects. In any event, isolated vacancies are unstable above 140°K in c-Si, and these should thus not be present in a-Si:H under ordinary conditions. [It is interesting that the isolated vacancy in c-Si is very likely characterized by a negative U_{eff} (Baraff *et al.,* 1980).] On the other hand, divacancies are stable in c-Si at room temperature and above, and it might appear that analogous defects could well be present in a-Si:H films, particularly after some hydrogen effusion has taken place. However, it should be borne in mind that a-Si:H, like all amorphous solids, is free from the constraints of periodicity, and atomic relaxation is thus considerably easier in an amorphous network than in a crystal. Consequently, it is far from clear that well-defined vacancy-type defects actually do exist in amorphous materials. In particular, the Si atoms surrounding the two different vacancies in a divacancy in c-Si are third-nearest neighbors. As pointed out previously, there is essentially no correlation between third neighbors in a-Si films (Moss and Graczyk, 1970), making it very unlikely that the structure of any existing divacancies in such films is well defined and reproducible.

On the other hand, it is well known that amorphous solids often contain extended defects such as cracks and voids, and there is structural evidence for these in a-Si (Barna *et al.,* 1977). Even in a-Si:H, columnar growth has been documented (Knights and Lujan, 1979), and thus it is likely that cracks exist between the columns. Surprisingly, NMR experiments (Reimer *et al.,* 1980) have suggested that a-Si:H films of high as well as low quality contain regions with relatively low hydrogen concentration (islands) interspersed in regions with considerably larger hydrogen concentration (tissues). Phillips (1979b) proposed that a-Si:H films form via the deposition of self-limiting clusters that do not fit together well, leaving voids within the material. The electronic structure of voids and cracks is controlled by the nature of the internal surfaces. Some insight into these could, in principle, come from studies of c-Si surfaces, but unfortunately the latter are complex and their detailed nature is still somewhat uncertain at present. It is known that all surfaces of c-Si reconstruct, and the (lll) surface reconstruction is such that neighboring dangling bonds pair to eliminate any unpaired spins. Since this is the only surface containing single dangling bonds, we should expect an amorphous analogue to predominate on the internal surfaces of voids in a-Si:H. Phillips (1979b) suggested that internal edges exist in a-Si films, and that after a pairing these yield the unpaired spins that are always observed in this material. This model accounts for the facts that (1) much more hydrogen can be introduced into a-Si than the values of N_s indicate and (2) much more hydrogen effuses at high temperatures than unpaired spins are created. However, these results are also consistent with the presence of isolated T_2^0 centers or T_3^+–T_3^- pairs. (In fact, the only important difference between ICTDs and paired dangling bonds is the presence of a dipole moment on the former.)

Surface states exist in the gap of c-Si, and we would thus expect that voids in a-Si:H also give rise to such states. Their detailed nature cannot be definitively estimated within our present state of knowledge, but their position in c-Si suggests that they may lie below midgap in a-Si:H. There is some experimental evidence for localized states in this region (Cohen *et al.*, 1980), although there are many possibilities for their origin.

IV. Localized States due to Impurities

13. PHOSPHORUS

Since phosphorus is the most common n-type dopant in a-Si:H, it is important to analyze the possible localized states arising from its presence. Since the strength of the Si–P bond is comparable to that of the Si–H bond (Adler, 1978) and P has an electronegativity only slightly lower than that of H, it might be expected that the bonding and antibonding orbitals arising from normal structural Si–P bonding have roughly the same energies as those from Si–H bonds; that is, the bonding orbitals lie well inside the valence band, but the antibonding orbitals are located near the conduction-band mobility edge. Normal structural bonding for phosphorus (a pnictogen atom, hereinafter designed P is our notation scheme) is P_3^0 (p^3 bonding). If all P atoms introduced into a-Si:H entered in P_3^0 configurations, P would not act as a dopant but would instead play a similar role to that of H, namely, as a network relaxer and defect compensator that also tends to open up the band gap. Clearly, this is not the case: the introduction of P into a-Si:H increases E_F from the midgap region to a position about 0.1 eV from the conduction-band mobility edge (Magariño *et al.*, 1982). Thus it is very likely that a significant fraction of the P atoms enter the a-Si:H network in a tetrahedral (sp^3) configuration. It is difficult to determine the ratio of P_4 to P_3 centers in P-doped a-Si:H accurately, but a typical estimate is about 30% (LeComber and Spear, 1979). Since P_3^0 should be the lowest-energy configuration and such centers also introduce less strain into the network than P_4 centers, the doping of a-Si:H by P presents a major puzzle. The resolution may be the low energy of P_4^+–T_3^- pairs (Adler, 1980d). In fact, since the Si–P bond is stronger than the Si–Si bond, a P_4^+–T_3^- pair may have lower energy than a P_3^0–T_4^0 pair; the total numbers of both hybridized (4) and unhybridized (3) bonds are preserved, but (assuming that all bonds are to the predominant Si atoms) a P_4^+–T_3^- pair contains one more Si–P bond than a P_3^0–T_4^0 pair. In fact, if this is the case, the P_3^0 centers should be considered as defects, introduced to relieve network strains. The local configurations of P_4^+ and P_3^0 centers are sketched in Figs. 6a and 6b, respectively.

The electronic structure arising from the presence of P_4^+–T_3^- pairs depends critically on the sign of U_{eff} for T_3 defects. If $U_{\text{eff}} > 0$, the band structure for low and moderate P concentrations is as sketched in Figs. 3b

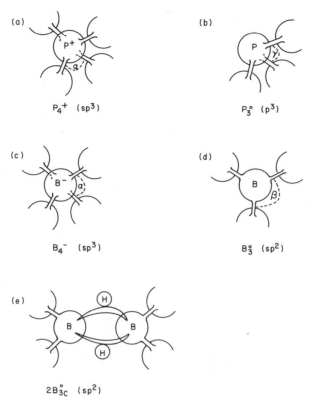

FIG. 6. Sketches of the optimal local coordinations of dopant atoms in several different configurations: (a) positively charged tetrahedrally coordinated P atom, P_4^+; (b) neutral three-fold-coordinated P atom, P_3^0; (c) negatively charged tetrahedrally coordinated B atom, B_4^-; (d) neutral three fold-coordinated B atom, B_3^0; (e) complex consisting of two B three-center bonds with bridging H atoms. The bond angles are the same as in Fig. 2: $\alpha = 109.5°$, $\beta = 120°$, $\gamma \approx 95°$.

and 3c, respectively. E_F rises continuously to a point about halfway between the P_4^+ and T_3^- bands. If, on the other hand, $U_{eff} < 0$, the band structure is as sketched in Figs. 4b and 4c. In this case, E_F is pinned until the P_4^+ concentration becomes equal to the concentration of Si dangling bonds; only for greater P_4^+ concentrations does E_F increase. It is difficult to ascertain the sign of U_{eff} from the available data on P-doped a-Si:H. In high-quality samples the Si dangling-bond concentration is very low ($\sim 10^{16}$ cm^{-3}), so that E_F would move up with even 1 ppm P even if $U_{eff} < 0$. The fact that E_F increases to within about 0.1 eV of the conduction-band mobility edge appears to favor the $U_{eff} > 0$ model (since it requires higher energies for the T_3^- centers). However, this conclusion becomes less certain when the ESR data are taken into account. As P is introduced, the concentration of T_3^0

centers decreases, as must be the case when E_F increases (thus converting the T_3^0 sites into spin-paired T_3^- centers). With the introduction of more P, the T_3^0 spin signal is completely quenched and a new ESR signal is found with $g = 2.004$ and a linewidth of about 10 G (Magariño *et al.,* 1982). These unpaired spins could arise from P centers such as P_4^0 or P_3^-; however, the fact that the same signal is observed in light-induced ESR in undoped a-Si:H (Street and Biegelsen, 1980) eliminates this possibility. The conventional model is that these spins arise from electrons in the conduction-band tail (i.e., T_4^- centers). However, there are several difficulties with this interpretation. The ESR signal maintains the same g value and linewidth despite the wide array of bond-angle strains ($\pm 10°$) expected to give rise to states in the band tail. Furthermore, as the P concentration increases, the ESR signal goes through a maximum and begins to decline in strength (Friederich and Kaplan, 1978; Magariño *et al.,* 1982). This is inconsistent with motion of E_F through a band tail in which $g(E)$ is expected to increase monotonically with energy. In addition, E_F increases with increasing temperature (Hasegawa *et al.,* 1980), a result also incompatible with a rising $g(E)$ with energy. A simple resolution (Adler, 1981) is that this ESR signal arises from T_2^- centers. Such centers should be localized in energy, so tht $g(E)$ is not monotonically increasing. The quenching of the signal is then easily explained by the formation of T_2^{2-} centers with increasing P concentration. This model also explains the increase of E_F to within 0.1 eV of the conduction-band mobility edge, since the presence of $2P_4^+ - T_2^{2-}$ pairs results in an E_F between the T_2^{2-} and P_4^+ levels, both of which must be very near E_c.

Another possible configuration for pnictogen atoms is as a P_2^- center (Adler, 1980c). This most easily could appear as part of a $T_3^+ - P_2^-$ pair. However, it should be noted that such a pair possesses two fewer bonds than a $P_4^+ - T_3^-$ pair without any reduction in s-p promotion energy, and thus it is energetically unfavorable. Nevertheless, we cannot eliminate the possibility of formation of P_2^- centers, particularly as a strain-relieving mechanism.

To summarize, the introduction of P (or any other column V dopant) could lead to the presence of $P_4^+ - T_3^-$ pairs, P_3^0 centers, $P_4^+ - T_2^-$ pairs, $2P_4^+ - T_2^{2-}$ triplets, or $T_3^+ - P_2^-$ pairs. It is probably that the first of these represents normal structural bonding and is responsible for at least light and moderate *n*-type doping. It should again be noted that the oppositely charged defects could be spatially close, changing their resulting energy levels in an analogous manner to the intrinsic defects discussed in Part III.

14. BORON

In principle, the centers and their electronic structure arisng from the introduction of B, the most common *p*-type dopant, into a-Si:H should be analogous to those resulting from P. Unfortunately, becase of the complex chemistry of boron, this is not the case.

The introduction of B into a-Si:H lowers the position of E_F, and high concentrations reduce E_F to about 0.1 eV above the valence-band mobility edge. The spin signal arising from the T_3^0 centers is quenched by B in a similar manner to P, and a new ESR signal appears, with $g = 2.013$ and a linewidth of about 15 G (Magariño *et al.*, 1982). This signal also goes through a maximum with increasing B concentration (Hasegawa *et al.*, 1980). All of these results are analogous to those for P doping, and we can conclude that it is likely $T_3^+-B_4^-$ (Fig. 6c) pairs form in a-Si:H:B films. However, it is not clear whether or not this pair or B_3^0 centers (Fig. 6d) represents normal structural bonding in a-Si:H:B. Bond dissociation energies (Kondratiev, 1974) indicate that the Si–B bond might be weaker than the Si–Si bond, which would favor the formation of B_3^0 centers. There is also evidence from NMR data (Carlos *et al.*, 1982) that the vast majority of the boron in a-Si:H:B is three-fold-coordinated. Since B is more electronegative than Si, and B_3^0 centers have two empty nonbonding p orbitals, the presence of B_3^0 may introduce states into the gap of a-Si:H:B.

The electronic structure of the active $T_3^+-B_4^-$ pairs depends on the sign of U_{eff}. If $U_{eff} > 0$, the band structure of lightly and moderately B-doped a-Si:H films is as sketched in Figs. 3d and 3e, respectively, while the analogous results for $U_{eff} < 0$ are shown in Figs. 4d and 4e. As is the case for P doping, heavy B doping requires the formation of $T_2^+-B_4^-$ pairs and ultimately $T_2^{2+}-2B_4^-$ triplets. The existence of these is also consistent with the association of T_2^+ centers with the $g = 2.013$ ESR signal and the observed maximum in that signal with increasing B concentration (Adler, 1981).

A more unlikely possibility, but one that should not be overlooked completely is a B_2^+ center, perhaps as part of a $B_2^+-T_3^-$ pair. Such a pair is very energetically unfavorable, possessing two fewer bonds than either a $T_4^0-B_3^0$ or a $T_3^+-B_4^-$ pair, but it does represent an efficient strain-relieving mechanism.

Boron chemistry allows for an interesting new possibility, the formation of three-center bonds with bridging hydrogen atoms (Ovshinsky 1977; Ovshinsky and Adler, 1978), B_{3c}^0. A defect complex involving two such three-center bonds is sketched in Fig. 6e. This type of a $2B_{3c}^0$ complex saturated by four covalently bonded hydrogen atoms is the chemical basis for the B_2H_6 molecule, the gas of which is universally used to dope a-Si:H p-type. A $B_{3c}^0-T_{3c}^0$ complex is also possible in a-Si:H:B. The electronic structure of a $2B_{3c}^0$ complex is such that empty nonbonding orbitals are introduced into the gap. The relatively large electronegativity of B suggests that these states may fall in the lower part of the gap, and thus they can account for p-type doping without the formation of B_4^- centers. As yet, there is no direct experimental evidence for large concentrations of three-center bonds in a-Si:H:B films, but the possibility must still be borne in mind.

Finally, it should be noted that the dissociation energy is slightly greater for B–B than for B–Si bonds (Kondratiev, 1974). This suggests the possibility of boron clustering in a-Si:H:B, even at relatively low B concentrations. Such clustering is consistent with the NMR data (Carlos *et al.,* 1982).

15. OXYGEN

It is extremely difficult, even under high-vacuum conditions, to prepare a-Si:H films without the incorporation of some oxygen; typically O concentrations of the order of 10^{20} cm^{-3} are observed. Oxygen is the second most electronegative atom in the Periodic Table and there is a strong ionic component to all of its bonds. It forms very strong bonds with both Si (approximately 4.0 eV) and H (5.0 eV).

Since oxygen is a chalcogen atom (hereinafter labeled C in our notation), from column VI in the Periodic Table, normal structural bonding is C_2^0. In general, as described by Kastner *et al.* (1976), chalcogen atoms in amorphous networks possess a low-energy defect, a $C_3^+-C_1^-$ pair. Such a pair has been suggested for a-SiO$_2$ (Lucovsky, 1979). In principle, the creation energy of a $C_3^+-C_1^-$ pair in a-Si:H:O could be sufficiently small that a sizable thermodynamic concentration of $C_3^+-C_1^-$ pairs should always be present. However, the saturated nature of the T_4^0 centers together with the large oxygen electronegativity serves to effectively increase the creation energy of $C_3^+-C_1^-$ pairs in a-Si:H:O, and they probably do not appear in sufficiently large concentrations to significantly affect the electronic properties of the film.

The actual lowest-energy defect in a-Si:H:O is most likely a neighboring $C_3^0-T_3^0$ pair. In this case the extra bond is a dative bond between the lone pair on the O atom and the empty p orbital on the dehybridized Si atom (in an s^2p^2 configuration). Such pairs are very likely characterized by a negative U_{eff} (Adler, 1980c), and thus if present in significant concentrations they would tend to pin E_F. A $T_3^+-C_1^-$ pair has a much larger creation energy, but it represents an effective strain-relieving mechanism. C_1^- centers are called nonbridging and have been identified as common defects in a-SiO$_2$ films.

The strength of the O–H bond together with the ubiquitous presence of H$_2$O in the environment during film preparation suggest that OH groups exist in a-Si:H:O. These bond monovalently and would not then be expected to give rise to any states in the gap. However, their existence leads to the possibility of additional weak hydrogen bonding, for example, O–H$\cdots$O bonds. Such bonds could play an important role in nonequilibrium phenomena, since they are easily broken.

16. OTHER IMPURITIES

Other important impurities in a-Si:H, introduced either accidentally or intentionally, include N, C, and halogens. The most significant of the halogens is fluorine, which has been used to relieve strains and compensate defects (Ovshinsky and Madan, 1978). Fluorine is the most electronegative element of all and since ionic bonds do not constrain bond angles it very likely provides a better strain-relieving mechanism than hydrogen. The Si–F bond strength (about 6.2 eV) is so strong that it is extremely unlikely that the presence of F introduces any states in the gap of a-Si:H:F. Furthermore, F atoms are very unlikely to appear in any defect configuration.

V. Localized States near Surfaces and Interfaces

17. SURFACES

In general, free surfaces act similarly to the internal surfaces on voids discussed in Part III. Cooperative reconstructions can occur analogous to those that characterize surfaces of c-Si. These almost certainly introduce localized surface states into the gap. Since the position of E_F in the bulk is very unlikely to match that near the surface, space-charge effects are very important near surfaces. In addition, H, O, or OH can bond to potential dangling bonds at free surfaces, depending on the film preparation technique and its subsequent history. Guha et al. (1980) measured the field effect on a free surface of a-Si:H and observed both space-charge effects and a sensitivity to annealing.

18. INTERFACE STATES

The most important interfaces for a-Si:H films are those near SiO_2 regions used for isolation and those near metallic regions used for electrical contacts. Clearly, space-charge effects are extremely important near these surfaces. DLTS measurements, which can, in principle, distinguish bulk from interface states, have provided evidence for enhanced concentrations of localized gap states near such interfaces (Cohen et al., 1980). The present evidence is that the interface charge density near a-Si:H–a-SiO_2 interfaces is less than 10^{11} cm^{-2} (Goodman and Fritzsche, 1980; Williams et al., 1979). Schottky barriers near a-Si:H–metal interfaces are characterized by considerably larger values of charge density (Dalal, 1980). Transition-metal atoms that enter a-Si:H can be expected to introduce large concentrations of localized gap states (Ovshinsky and Adler, 1978).

VI. Nonequilibrium Effects and Metastable States

19. TRANSIENT EFFECTS

In a nonequilibrium experiment large concentration of excess free carriers are created by, for example, photogeneration, injection, and so on. The driving force (e.g., light, electric field) is applied for a certain time and then removed. Specific physical properties such as conductivity, optical absorption, luminescence, and so on, are then monitored as functions of time, both in the presence of the driving force and as the system returns to equilibrium. Once any nonequilibrium conditions are obtained, the concepts of a unique E_F and a well-defined statistical probability of occupancy, $f(E)$, must be modified. Instead, the localized states act as carrier traps and recombination centers, with kinetics determned by their populations, states of occupancy, and cross sections. Rose (1978) simplified the complex problem by introducing the concept of *demarcation levels* for trapped carriers: for example, electrons trapped above an energy, E_{Fn}, are more likely to be reemitted into the conduction band than to recombine with a hole below E_F, while the reverse is true for electrons trapped below E_{Fn}; the hole demarcation level E_{Fp} is defined analogously. Once all the excess free carriers have recombined, the system has returned to equilibrium.

The main problems in the analysis of transient phenomena arise from the effects of electronic correlations and local relaxations. When correlations are important, there are necessary changes in $g(E)$ with occupancy (Adler, 1982d). Under ordinary conditions these are controlled by the value of U_{eff} for the predominant defect.

If $U_{eff} > 0$, the density of states at equilibrium is given by one of the curves in Fig. 7, depending on the concentration of electrons in the defect levels n, where n is defined as the average number of electrons per defect site. In p-type material E_F lies in the lower defect band, as in Fig. 7a, and the majority of electron traps are located just above E_F; since these states are positively charged, they have a relatively high cross section and their location suggests that they act as recombination centers. In addition, the process of trapping shifts states from the lower to the upper defect band, converting the charged recombination centers to neutral electron traps. Most of the hole traps are neutral and located just below E_F. In n-type samples E_F lies in the upper defect band (see Fig. 7c), and the negatively charged hole recombination centers dominate the reequilibration kinetics.

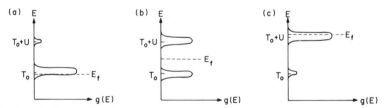

FIG. 7. Effective density of one-electron states $g(E)$ at equilibrium as a function of energy E for a defect center with a positive effective correlation energy $U > 0$: (a) $n < 1$, (b) $n = 1$, (c) $n > 1$, where n is the average number of electrons per defect site.

For intrinsic samples (see Fig. 2b), there are roughly equal densities of neutral electron traps well above E_F and hole traps well below E_F. Although the process of carrier trapping shifts states between the defect bands, the overall density of states does not change significantly because both electrons and holes are trapped with approximately equal rates. However, since this trapping creates a charged recombination center for the opposite carrier type, bimolecular recombination kinetics should predominate.

For $U_{eff} < 0$, the situation is qualitatively different. The equilibrium density of states (ignoring any spatial correlations between the charged defects) is given by one of the sketches in Fig. 8. Since isolated negatively correlated defects pin E_F to the extent of their concentration, the material is intrinsic until it is doped at levels greater than those necessary to force the defects to be either entirely positively charged (e.g., T_3^+) or entirely negatively charged (e.g., T_3^-). However, for any intermediate condition of occupancy $(0 < n < 2)$, the electron traps are positively charged and located well above E_F and the hole traps are negatively charged and located well below E_F. The interesting feature is that upon either a positively charged center trapping an electron or a negatively charged center trapping a hole, the subsequent local relaxation causes both defects to become *identical* neutral centers (e.g., T_3^0); these neutral defects are located in different positions in energy than were the charged centers, thus leading to a $g(E)$ such as either of those sketched in Fig. 9. In the absence of the relaxation, trapping ordinarily would not be expected to be a recombination process, since the probability of reemission is likely to be larger than that of trapping of the opposite carrier type by the neutral centers. However, the relaxation process lowers the energy of the trapped electrons to below E_F and raises the energy of the trapped holes above E_F. When these metastable centers trap carriers, a recombination process has taken place. The subsequent local relaxation then reestablishes equilibrium.

A major additional complication to the analysis of transient phenomena is the time scale of the relaxation processes. Ordinarily, such relaxations should take place within typical phonon times ($\sim 10^{-12}$ sec) causing them to

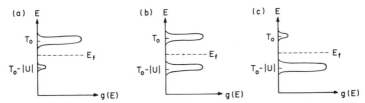

FIG. 8. Effective density of one-electron states $g(E)$ at equilibrium as a function of energy E for a defect center with a negative effective correlation energy $u < 0$: (a) $n < 1$, (b) $n = 1$, (c) $n > 1$.

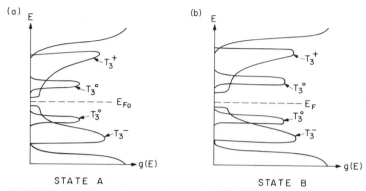

FIG. 9. Effective one-electron density of states for a-Si : H films under the assumption that dangling-bond defects are characterized by $U_{\text{eff}} < 0$ if complete atomic relaxations are possible, but that such relaxations are retarded in certain strained regions, resulting in the presence of stable T_3^0 centers: (a) equilibrium state A; (b) light-soaked state B.

be essentially instantaneous compared to the time scale of most transient experiments (see, however, Tauc, 1983). However, we must take into account the possible existence of potential barriers retarding the relaxations (Frye and Adler, 1981). Such barriers could slow down the relaxation processes by a factor of $\exp(E_b/kT)$, where E_b is the height of the barrier; for example, when $E_b = 0.4$ eV, the relaxation time at $300°$K becomes 10 μsec, while $E_b \simeq 1.0$ eV results in room-temperature relaxation times of the order of days. In the latter case nonequilibrium processes produce metastable states, and reequilibration at reasonable times then requires an annealing of the sample.

If a distribution of spatially correlated ICTDs is present, it becomes likely that the electron and hole demarcation levels cross the density of states resulting from these defects; thus some defect states represent traps and others recombination centers, even in the absence of rapid relaxations (Adler, 1982b).

20. METASTABLE STATES INDUCED BY EXCESS CARRIERS

Staebler and Wronski (1980) first observed that a flux of photons with energies greater than about 1.6 eV can produce a metastable state in a-Si : H films. They called the fully annealed (i.e., equilibrium) phase state A and the metastable phase state B. The A → B transition can be induced by a process that creates excess free carriers, but state A can be restored only by annealing. States A and B are characterized by very diverse physical properties (Adler, 1983; Schade, 1983).

Many models have been proposed to account for the A → B transition, ordinarily attributing it to either the creation of new defects or the interconversion of existing defects by the trapping or recombination of the excess carriers. However, the previous discussion suggests a natural interpretation in terms of defects characterized by a negative U_{eff}, with a potential barrier retarding local relaxations. A specific model (Adler, 1983) identifies the responsible defect as the Si dangling bond. The hypothesis is that $U_{eff} < 0$ for this defect provided that complete local relaxations can occur around both the T_3^+ and T_3^- centers in the network, but because of the overconstrained nature of a-Si:H films large concentrations of isolated T_3^0 centers (characterized by positive values of U_{eff}) are simultaneously present in all samples. The resulting $g(E)$ is sketched in Fig. 9a. Upon the trapping of excess electrons by T_3^+ centers or holes by T_3^- centers, additional T_3^0 centers are produced after the appropriate relaxations. This yields state B, for which $g(E)$ is as sketched in Fig. 9b. If there were no significant potential barrier retarding the reequilibration process, $2T_3^0 \rightarrow T_3^+ + T_3^-$, then state B would be unstable. However, this process requires two bond-angle distortions, effectively changing the two 109.5° angles that characterize the T_3^0 centers to one of 120° (T_3^+) and one near 95° (T_3^-). A possible configuration-coordinate diagram for this reequilibration process is given in Fig. 10. A large value of E_b for such a process would not be surprising. Clearly, this is only one possible explanation for an effect that could have many possible origins, depending on the details of the preparation conditions and the impurity concentrations in the particular film being investigated.

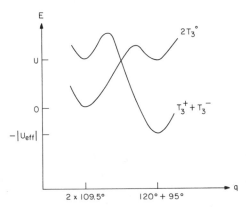

FIG. 10. Total energy as a function of a configuration coordinate q that represents the variations in both of the sets of bond angles around a pair of dangling-bond defects. Two charge configurations are possible for the pair, $2T_3^0$ and $T_3^+ + T_3^-$, and the energies of both are plotted. The two curves shown will interact, resulting in a potential barrier between the lowest-energy states of the possible charge configurations.

VIII. Conclusions

This chapter has been primarily concerned with the origin, nature, and density of states of the defects that characterize films of hydrogenated amorphous silicon. The principal features of hydrogenated amorphous silicon are (1) the absence of low-energy defects, leading to only small concentrations of thermodynamically induced defects, and (2) the overconstrained nature of the network, resulting in the presence of significant concentrations of strain-related defects. Both of these features arise because of the tetrahedral bonding in the films. The fact that the defects are induced by strains rather than arising from thermodynamic requirements precludes a simple analysis of their nature and leads to many potential complications, including a diversity of physical properties depending on the details of the preparation conditions and film history. Nevertheless, at this point in our understanding of the material, there is a glimmer of hope that the mystery is beginning to unravel. The biggest need at present is more rigorous theoretical calculations. These have lagged because of the intrinsic difficulties in simplifying the problem of nonperiodic networks in general taken together with the enormous complexities arising from the importance of electronic correlations and electron–phonon interactions. For these same reasons, it is dangerous to blindly accept the conclusions of calculations that completely neglect these effects.

ACKNOWLEDGMENT

The research described in this chapter was supported by the National Science Foundation Materials Research Laboratory Grant No. DMR 81-19295.

REFERENCES

Adler, D. (1978). *Phys. Rev. Lett.* **41**, 1755.
Adler, D. (1980a). *J. Chem. Educ.* **57**, 560.
Adler, D. (1980b). *Sol. Cells* **2**, 199.
Adler, D. (1980c). *J. Non-Cryst. Solids* **35–36**, 819.
Adler, D. (1980d). *J. Non-Cryst. Solids* **42**, 315.
Adler, D. (1981). *J. Phys. (Orsay, Fr.)* **42**, C4-3.
Adler, D. (1982a). *J. Solid State Chem.* **45**, 40.
Adler, D. (1982b). *Sol. Energy Mater.* **8**, 53.
Adler, D. (1982c). *Naturwissenschaften* **69**, 574.
Adler, D. (1982d). *In* "Handbook on Semiconductors" (T. S. Moss, ed.), Vol. 1, pp. 805–841. North-Holland Publ., Amsterdam.
Adler, D. (1983). *Sol. Cells* **9**, 133.
Adler, D., and Yoffa, E. J. (1976). *Phys. Rev. Lett.* **36**, 1197.
Adler, D., and Yoffa, E. J. (1977). *Can. J. Chem.* **55**, 1920.

Allen, D. C., and Joannopoulos, J. D. (1980). *Phys. Rev. Lett.* **44**, 43.

Anderson, P. W. (1975). *Phys. Rev. Lett.* **34**, 953.

Baraff, G. A., Kane, E. O., and Schlüter, M. (1980). *Phys. Rev. B* **21**, 3563.

Barna, A., Barna, P. B., Radnoczi, G., Toth, L., and Thomas, P. (1977). *Phys. Status Solidi A* **41**, 81.

Biegelsen, D., Street, R. A., Tsai, C. C., and Knights, J. C. (1979). *Phys. Rev. B* **20**, 4839.

Carlos, W. E., and Taylor, P. C. (1982). *Phys. Rev. B* **25**, 1435.

Carlos, W. E., Greenbaum, S. G., and Taylor, P. C. (1982). *Bull. Am. Phys. Soc.* [2] **27**, 208.

Chakraverty, B. K. (1981). *J. Phys. (Orsay, Fr.)* **42**, C4-749.

Ching, W. Y., Lam, D. J., and Lin, C. C. (1979). *Phys. Rev. Lett.* **42**, 805.

Ching, W. Y., Lam, D. J., and Lin, C. C. (1980). *Phys. Rev. B* **21**, 2378.

Cody, G. D., Wronski, C. R., Abeles, B., Stephens, R. B., and Brooks, B. (1980). *Sol. Cells* **2**, 227.

Cohen, J. D., Lang, D. V., and Harbison, J. P. (1980). *Phys. Rev. Lett.* **45**, 197.

Dalal, V. L. (1980). *Sol. Cells* **2**, 261.

Dersch, H., Stuke, J., and Beichler, J. (1981). *Phys. Status Solidi B* **105**, 265.

Eberhart, M. E., Johnson, K. H., and Adler, D. (1982). *Phys. Rev. B* **26**, 3138.

Elliott, S. R. (1978). *Philos. Mag. [Part] B* **38**, 325.

Friederich, A., and Kaplan, D. (1978). *J. Electron. Mater.* **7**, 253.

Fröhlich, H. (1954). *Adv. Phys.* **3**, 325.

Frye, R. C., and Adler, D. (1981). *Phys. Rev. B* **24**, 5485.

Goodman, N. B., and Fritzsche, H. (1980). *Philos. Mag. [Part] B* **42**, 149.

Guha, S., Narasinhan, K. L., Navkhandewela, R. V., and Pietrussko, S. M. (1980). *Appl. Phys. Lett.* **37**, 572.

Hasegawa, S., Shimiza, T., and Hirose, M. (1980). *J. Phys. Soc. Jpn.* **49**, *Suppl.* A, 1237.

Hirose, M. (1981). *J. Phys. (Orsay, Fr.)* **42**, C4-705.

Holstein, T. (1959). *Ann. Phys. (N.Y.)* **8**, 325, 353.

Joannopoulos, J. D. (1980). *J. Non-Cryst. Solids* 35–36, 781.

Johnson, K. H., Kolari, H. J., de Neufville, J. D., and Morel, D. L. (1980). *Phys. Rev. B* **21**, 643.

Kastner, M., Adler, D., and Fritzsche, H. (1976). *Phys. Rev. Lett.* **37**, 1504.

Knights, J. C., and Lujan, R. (1979). *Appl. Phys. Lett.* **35**, 244.

Kondratiev, V. N. (1974). "Bond Dissociation Energies, Ionization Potentials and Electron Affinities." Nauka, Moscow.

LeComber, P. G., and Spear, W. E. (1979). *In* "Amorphous Semiconductors" (M. H. Brodsky, ed.), p. 251. Springer-Verlag, Berlin and New York.

Lucovsky, G. (1979). *Philos. Mag.* **B39**, 513.

Magariño, J., Kaplan, D., Friederich, A., and Deneuville, A. (1982). *Philos. Mag. [Part] B* **45**, 285.

Moss, S. C., and Graczyk, J. F. (1970). *Proc. Int. Conf. Phys. Semicond., 10th, 1970* p. 658.

Mott, N. F., and Davis, E. A. (1979). "Electronic Processes in Non-Crystalline Materials," 2nd ed. Oxford Univ. Press (Clarendon), London and New York.

Moustakas, T. D., Anderson, D. A., and Paul, W. (1977). *Solid State Commun.* **23**, 155.

Ovshinsky, S. R. (1977). *Amorphous Liq. Semicond., Proc. Int. Conf., 7th, 1977* p. 519.

Ovshinsky, S. R., and Adler, D. (1978). *Contemp. Phys.* **19**, 109.

Ovshinsky, S. R., and Madan, A. (1978). *Nature (London)* **276**, 482.

Papaconstatopoulos, D. A., and Economou, E. N. (1981). *Phys. Rev. B* **24**, 7233.

Phillips, J. C. (1979a). *J. Non-Cryst. Solids* **34**, 153.

Phillips, J. C. (1979b). *Phys. Rev. Lett.* **42**, 151.

Reimer, J. A., Vaughan, R. W., and Knights, J. C. (1980). *Phys. Rev. Lett.* **44**, 193.

Rose, A. (1978). "Concepts in Photoconductivity and Allied Problems." Krieger, Huntington, New York.

Schade, H. (1984). *In* "Semiconductors and Semimetals" (R. K. Willardson and A. C. Beer, eds.), Vol. 21B, Chapter 11. Academic Press, New York.

Singh, J. (1981). *Phys. Rev. B* **23**, 4156.

Sol, N., Kaplan, D., Dieumegard, D., and Dubreuil, D. (1980). *J. Non-Cryst. Solids* **35–36**, 291.

Staebler, D. L., and Wronski, C. R. (1980). *J. Appl. Phys.* **51**, 3262.

Street, R. A., and Biegelsen, D. K. (1980). *J. Non-Cryst. Solids* **35–36**, 651.

Tauc, J. (1984). *In* "Semiconductors and Semimetals" (R. K. Willardson and A. C. Beer, eds.), Vol. 21B, Chapter 9. Academic Press, New York.

von Roedern, B., Ley, L., and Cardona, M. (1977). *Phys. Rev. Lett.* **39**, 1576.

von Roedern, B., Ley, L., Cardona, M., and Smith, F. W. (1980). *Philos. Mag [Part] B* **40**, 433.

Williams, R. H., Varma, R. R., Spear, W. E., and LeComber, P. G. (1979). *J. Phys. C* **12**, L209.

Index

319

Contents of Previous Volumes

Volume 20   **Semi-Insulating GaAs**